BEACHES OF NOVA SCOTIA
DISCOVERING THE SECRETS OF THE PROVINCE'S MOST BEAUTIFUL BEACHES

ALLAN BILLARD

PHOTOGRAPHY BY DONNA BARNETT

FORMAC PUBLISHING COMPANY LIMITED
HALIFAX

For Gabriel Billard

Copyright © 2015 by Allan Billard and Donna Barnett
All photos by Donna Barnett, except the following, which are from Shutterstock. Page: 13 (bottom), 15, 16 (bottom), 17, 21 (right), 27, 50–51 (top), 54 (left), 63 (bottom), 69 (right), 71, 76, 77 (top), 82, 97, 99 (bottom), 100 (bottom), 101 (bottom), 122 (bottom), 139 (right), 142 (bottom).

All rights reserved. No part of this book may be reproduced or transmitted in any form or by any means, electronic or mechanical, including photocopying, or by any information storage or retrieval system, without permission in writing from the publisher.

Formac Publishing Company Limited recognizes the support of the Province of Nova Scotia through Film and Creative Industries Nova Scotia. We are pleased to work in partnership with the agency to develop and promote our creative industries for the benefit of all Nova Scotians. We acknowledge the financial support of the Government of Canada through the Canada Book Fund for our publishing activities. We acknowledge the support of the Canada Council for the Arts for our publishing program.

Cover design: Meredith Bangay

Library and Archives Canada Cataloguing in Publication

Billard, Allan, 1949- author

    Beaches of Nova Scotia : discovering the secrets of some of the province's most beautiful beaches / Allan Billard and Donna Barnett.

Includes index.
ISBN 978-1-4595-0379-3 (pbk.)

    1. Beaches—Nova Scotia. 2. Beaches—Nova Scotia—Pictorial works. 3. Coasts—Nova Scotia. 4. Coasts--Nova Scotia—Pictorial works. I. Barnett, Donna, photographer II. Title.

GB459.6.B54 2015   551.45'7   C2015-900219-2

Formac Publishing Company Limited
5502 Atlantic Street
Halifax, Nova Scotia, Canada
B3H 1G4
www.formac.ca

Printed and bound in China

CONTENTS

INTRODUCTION   5
MAP   10

EASTERN SHORE
① TAYLOR HEAD   12
② MARTINIQUE   18
③ LAWRENCETOWN   24
④ RAINBOW HAVEN   30

SOUTH SHORE
⑤ CRYSTAL CRESCENT   36
⑥ QUEENSLAND   42
⑦ HIRTLES   48
⑧ RISSERS & CRESCENT   52
⑨ CARTERS   58
⑩ SAND HILLS   62
⑪ THE HAWK & CAPE SABLE   66

BAY OF FUNDY
⑫ MAVILLETTE   72
⑬ BLOMIDON   78
⑭ BLUE BEACH   84
⑮ BURNTCOAT HEAD   90

NORTHUMBERLAND SHORE
⑯ BLUE SEA   94
⑰ RUSHTONS   98
⑱ MELMERBY   102
⑲ POMQUET   108
⑳ BAYFIELD   114

CAPE BRETON
㉑ PORT HOOD   120
㉒ WEST MABOU   124
㉓ INVERNESS   130
㉔ ASPY BAY   136
㉕ NORTH BAY   140
㉖ INGONISH   146
㉗ PONDVILLE   150

BEACHES ACT   154
ACKNOWLEDGEMENTS   155
INDEX   156

INTRODUCTION

WHAT FOLLOWS IS AN EXPOSÉ OF TWENTY-SEVEN well-known Nova Scotia beaches: their most attractive features and the characteristics that make them different from any other beaches.

That is the point; each beach is different. Walking on Martinique Beach on the Eastern Shore is very different from walking on Blomidon Beach on the Bay of Fundy. Martinique supplies a time-lapsed photo essay of storm-tossed sand and the here-today-gone-tomorrow nature of dune creation, while Blomidon offers a look back millions of years into a time when the continents were joined and dinosaurs roamed this land.

It may be best, however, to start with recent history.

In the late 1950s, the Macdonald Bridge across Halifax Harbour was newly opened and the urban renewal culture was spreading across the land. Many young families were moving to Dartmouth and building new homes there. For them, Silver Sands Beach at Cow Bay, just beyond Eastern Passage, was an added incentive to cross the harbour and live in a bedroom community. It was an easy drive from Dartmouth's growing residential communities. The privately owned strip of seacoast boasted more than a thousand metres of fine, white sand. In the summer season, a warm, brackish-water lake shimmered peacefully behind a small, treed campground. Cow Bay also became known for its delightful concrete animal sculptures created from the sands found on the site. Church groups held their annual Sunday school picnics there and (even before the fast food craze took hold) it was the best place to take the kids for ice cream and french fries. For others, the dance hall was an even bigger attraction. Silver Sands was a magnet for Dartmouthians and a gold mine for the landowners.

In Halifax the post-war boom was taking hold and the economy was bustling. White-collar jobs were replacing the wartime factory

BEACHES OF NOVA SCOTIA

INTRODUCTION

work and modern office towers were needed. Fine new structures were planned to replace the tired old inner city. Halifax City Council eagerly approved the expropriation of several downtown neighbourhoods and invited developers to bid on renewal projects. In 1966, a consortium of well-known local businessmen won the right from City Hall to build Scotia Square. Three new towers would be built on top of a modern new shopping centre, offering "Class A" office space and trendy fashion outlets. Plenty of parking and a modern traffic interchange was also part of the plan.

Residents of Dartmouth and Halifax went along with the urban renewal boosterism, but didn't think about where the building materials were to come from. Eager developers in Halifax needed unlimited concrete for Scotia Square — not to mention the city's first container terminal, which was also on the drawing boards. Those modern structures were tremendous opportunities for the region, and the closest source of construction sand for the needed concrete was the beach at Cow Bay. It wasn't long before the heavy equipment moved in and the beach was loaded into dump trucks. From there, the sand became concrete, the concrete became foundations and Halifax had its modern downtown centre. The sand at Silver Sands was gone in less than one season.

Today, the huge moose sculpture still stands at the entrance road to the beach, but it is the last of that era. Cow Bay is a "sometime" destination for hardcore surfers, but its shore is rocky and unfriendly. On the positive side, the sacrifice of Silver Sands was not lost on the provincial government. In 1989, the government of the day enacted the Beaches Act, which offered (for the first time) protection to the remaining provincial beaches. Most importantly, it restricted the removal of sand and other aggregates.

The embarrassing loss of Silver Sands was also the incentive to acquire Rainbow Haven Beach from private owners. The protected beach there has become the most popular in the province, attracting over twenty-five thousand enthusiastic visitors each summer and

BEACHES OF NOVA SCOTIA

INTRODUCTION

into the fall. It has proven to be an honest effort by government to atone for earlier mistakes.

There are now almost one hundred protected beaches in the province, many of them in fully developed provincial parks. Each of them, and its associated dune system, is recognized as a significant and sensitive environment. They are also a highly valuable recreational resource. The full range of land-use activities on Nova Scotia's beaches is regulated, including a prohibition on motor vehicles and a ban on selling things like T-shirts and other trinkets. Anything that might be considered an undesirable impact on a beach and its dunes is not allowed.

Beaches are considered mirrors, however, reflecting the way we treat our surroundings. They illustrate the way our society deals with vulnerable frontier land. A natural beach is held to be more valuable than a development site. As much as these beaches should be forever accessible to eager visitors, it is important to accept that natural disasters will occur, and when they do, nature can be relied upon to repair itself, in the fullness of time.

The following chapters and the sites they illustrate are the stories of some of Nova Scotia's best-known beaches and the reasons why they reflect the very best we can offer: to preserve and protect nature. If beaches are nature's frontiers, which by definition means they can be destroyed by nature as easily as they were created, then let it be. In the meantime, let them be free and open to the full enjoyment of all Nova Scotians and the visitors who are welcomed here.

Allan Billard
Donna Barnett

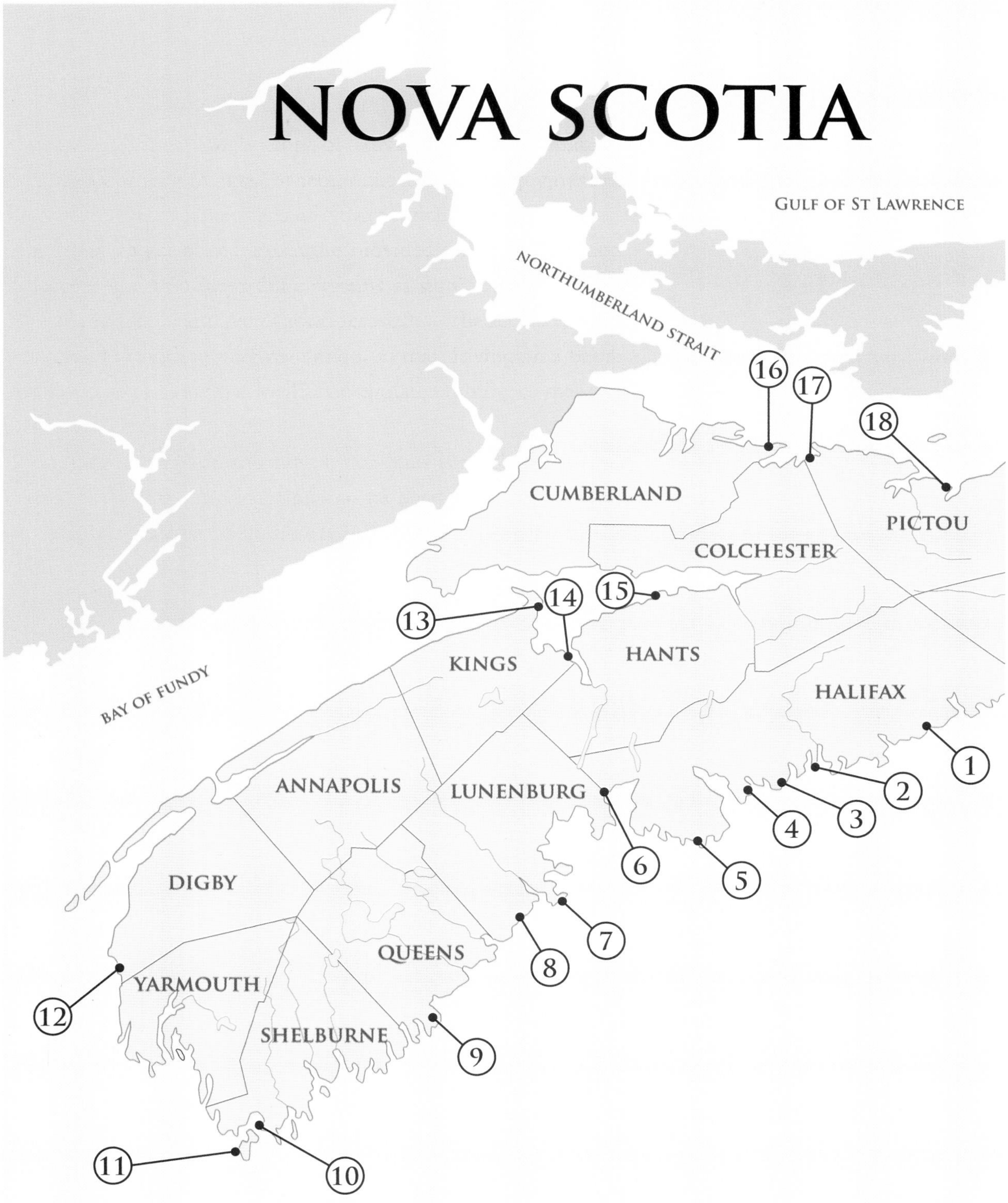
NOVA SCOTIA
GULF OF ST LAWRENCE
NORTHUMBERLAND STRAIT
BAY OF FUNDY
CUMBERLAND
COLCHESTER
PICTOU
KINGS
HANTS
HALIFAX
ANNAPOLIS
LUNENBURG
DIGBY
QUEENS
YARMOUTH
SHELBURNE
1
2
3
4
5
6
7
8
9
10
11
12
13
14
15
16
17
18

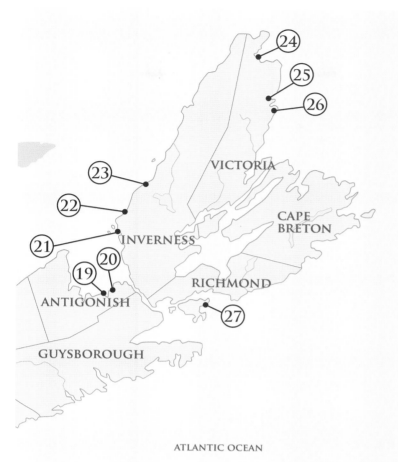

BEACHES OF NOVA SCOTIA LEGEND

① Taylor Head
② Martinique
③ Lawrencetown
④ Rainbow Haven
⑤ Crystal Crescent
⑥ Queensland & Cleveland
⑦ Hirtles
⑧ Rissers & Crescent
⑨ Carters
⑩ Sand Hills
⑪ The Hawk & Cape Sable
⑫ Mavillette
⑬ Blomidon
⑭ Blue Beach
⑮ Burntcoat Head
⑯ Blue Sea
⑰ Rushtons
⑱ Melmerby
⑲ Pomquet
⑳ Bayfield
㉑ Port Hood
㉒ West Mabou
㉓ Inverness
㉔ Aspy Bay
㉕ North Bay
㉖ Ingonish
㉗ Pondville

1

TAYLOR HEAD

Eastern Shore • 140 km east of Halifax

ALONG THE EASTERN SHORE ARE MANY lonely stretches of sea and sand. These idyllic coastal retreats, though, are most often visited by only the adventurous few who treasure the rugged wildness of coastal barrens like Taylor Head.

From this wind-swept scene, nestled into Psyche Cove, a solitary beach of silica white sand beacons like an oasis.

Within a sheltered backwater, surface currents spin off from the larger ocean swells, leaving fine sand on the beach. Even after thousands of years of carrying sands in from offshore, however, the beach has not accumulated enough deposits to completely cover its rocky foundation.

Standing alone on a featureless part of the beach, a solitary rock island becomes home for a community of thousands of tiny, shelled creatures and marine plants. The outgoing tide exposes many sea creatures that enjoy the plentiful sunlight but are still able to remain moist until the sea returns.

The park-access road leads right up to the beach where a few visitor amenities like change houses and vault toilets (a modern description for an outhouse) are provided. The natural features of the site, including the ancient rock formations and forests of balsam fir and spruce, are wonderfully explained on interpretative panels nearby.

When the first settlers were granted land along this coast, they chose Psyche Cove, with a view of the beach from their fields. Here, they harvested

ABOVE: *Immigrating United Empire Loyalists worked the rocky fields and meadows of this area as their first homesteads.*

LEFT: *The soft, white sand has migrated in from areas of the hard rock formations that make up the headland and coastal islands.*

a bounty of seafoods from the protected waters, which offered shelter to all — whether windborne or wave tossed. In modern times, this scene is left to the occasional young family or small group of hikers who enjoy the scenery from the original homestead sites.

Nesting just offshore, where many tiny islands offer protection for their young, thousands of migrating birds frequent the cove. Brazen gulls continuously survey the beach for the rare picnickers

Aggressive tides and currents torment the shoreline, eroding it and exposing the early geological epochs.

who might leave scraps for their next meal.

The common tern is a regular sight on these shores, feeding on schools of small fish, shrimp and other creatures that find their way into the shallow waters. Most days, though, there is nobody to appreciate their acrobatics as they wheel over the beach, capturing a silver-sided capelin to take back to their nestlings on the rocky offshore islands.

A number of different jellyfish species arrive from southern regions with the warm ocean currents, which can sometimes reach 20 degrees Celsius. Caught by a receding tide, many are stranded on the sand but can sometimes survive until the rising waters wash them back out to sea. What lies in wait for them back in the deeper water often proves more deadly than the sun-drenched beach.

Travelling up to forty nautical miles a day from their breeding grounds in the Caribbean, leatherback turtles follow the jellyfish migration and are known to gather at the feeding grounds near this beach. Resembling large seals, or logs in the water, they are not always correctly identified, but the

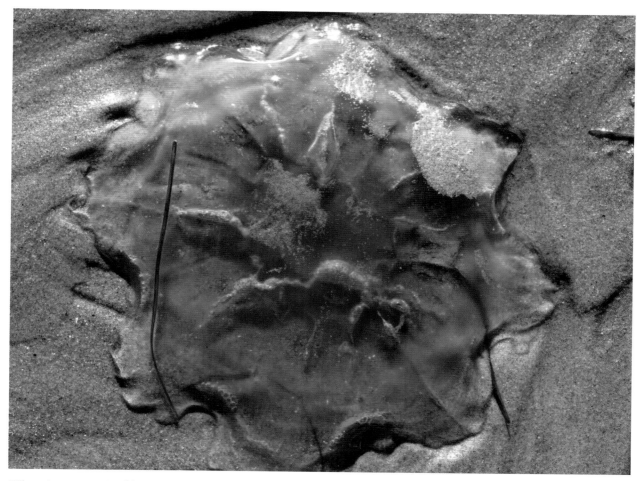

When the warmer Caribbean waters reach Nova Scotia in the fall, jellyfish rise to the surface to feed and breed. Many become stranded here and their decaying forms will nourish more fortunate beach dwellers and new elements of the food chain.

ABOVE: *A clump of samphire greens will migrate down the beach as far as it dares, loving the salt content of the sand, but not wanting to get its feet wet in the intertidal zone.*

RIGHT: *The common tern entertains visitors on this quiet shore with its acrobatic skill, snatching up the unsuspecting capelin in clear waters.*

number of positive sightings is increasing each season. It is likely that even more will be seen in the waters off Nova Scotia, where the visible increase in the population of jellyfish appears to be a good sign — at least for the leatherbacks.

This long peninsula is a splendidly natural site, an uncut diamond in the provincial park system. Taylor Head thrusts like a boney finger pointed directly out into the ocean winds and waves. With eighteen kilometres of hiking trails circling the headland, it

TAYLOR HEAD

Leatherback turtles follow the jellyfish and are often spotted along these shores.

offers some of the best views of the North Atlantic. Visit on a clear summer afternoon and the north-facing coastline will likely be sheltered from the sea breeze. On an overcast afternoon, the salty fog will smell of the vast ocean out where the breezes begin.

Most of the barrens are covered with a thin layer of dark humus and low shrubs. The limited soil barely covers the rock. It offers only enough depth for the hardy crowberry, clubmoss and dogwood to find root. There are also bogs, with hundreds of pitcher plants and patches of sparse forest, but they too suffer from the short summer and thin soil.

The brief summers at Taylor Head offer brisk winds, strong ocean currents and a very short growing season. When the warmth of a July afternoon raises the temperature of the top layer of sand, the wildness of the North Atlantic is forgotten and the beach is as inviting as any Caribbean getaway. For millennia, fish, birds, small and large plants and, until recently, homesteaders have sought out this sheltered cove of Taylor Head — ever the good provider and safe haven.

2

MARTINIQUE

Eastern Shore • 55 km east of Halifax

MARTINIQUE BEACH IS SAID TO BE the longest sandy beach in Nova Scotia. Actually, Aspy Bay in Cape Breton has it outdistanced if you add up all four pieces there, but this one stretch of golden sand can claim other bragging rights. It hosts a wildlife sanctuary, exceptional surfing and a good collection of sand dollars — treasured by children as souvenirs of their day at the beach.

Named after a long-forgotten sea battle in the Caribbean, this Martinique has become a favourite promenade for peaceful adventurers. It is a splendid venue to enjoy the mix of shorebirds, marine life, foxes and the flotsam and jetsam of the Atlantic Coast. With plentiful schools of fish nearby and rocky ledges to roost upon while they dry their wings, cormorants enjoy this shoreline too.

Never too far from the water's edge, or from their nests on the nearby dune, sandpipers seem in a constant frenzy. Stopping to dip their beaks into the sand, then scurrying off a few metres, then going airborne for no apparent reason, these sandpipers are feeding on those microscopic worms and shellfish that can only be caught when the water is less than a millimetre deep.

Farther up on the beach, the plant debris ripped from the seabed becomes nesting material for the sandpipers. Perhaps more importantly, the decaying salad of storm-tossed veggies will trap blowing sand, stabilize the area and offer a place for new tufts of marram grass to root. That in turn traps more sand,

ABOVE: *After major storms, the waves are less aggressive and often deposit new sands on the beach, but likely leave it further downstream. This is what keeps a beach "moving."*
LEFT: *When startled, first one sandpiper, then the whole flock, will flee farther down the beach, wheeling into the wind with frequent short glides.*

and a new dune ridge is able to grow.

Sand hurled up from one part of the beach by strong summer winds often becomes part of a high barrier dune. Martinique has a single spine, or dune ridge, that varies in height from "just taller than a man with a child on his shoulders" to perhaps twice that high in the middle. The ridge is typically steep on the water side and much more gradually sloping on the lee, or landward side. It acts like a bank account for the beach; dry sand gets deposited by high winds over time, but withdrawn and deposited

BEACHES OF NOVA SCOTIA

MARTINIQUE

ABOVE: *Cormorants are the ocean's vacuum cleaners. They are voracious feeders and the guano they leave behind is a valuable fertilizer for the beach, its plants and tiny animals.*

LEFT: *Sand hurled up the beach by strong summer winds becomes part of a high barrier dune that provides a place for grasses to gain a foothold.*

elsewhere as the beach moves, grows and rebuilds in damaged areas.

Harsh winter winds and the waves they drive scour sand from some dunes, leaving barely enough to anchor the grasses. However, the roots of marram grass are extremely long and hardy. They hold the sand against all but the strongest gales, drilling deep down into the dune's core and sending out runners to find trapped rainwater.

On summer days, children find the long green

The one long dune ridge on this beach has been breached by many Atlantic gales. What looks like a catastrophe for a few years soon starts to gather new plant life, however, and the creeping marram grasses slowly move in to restabilize the site.

spikes of marram perfect for sword fighting and other games, at least until they are reminded that the grass is an important part of the ecosystem, and should be left in place.

Most barrier dunes are a mixture of sand grains and larger materials, even glacial till and gravel. The Martinique dune is different; it is almost entirely fine sand, less stable than many along the coast and more easily breached by autumn gales.

Martinique Beach always seems to be on the brink of catastrophe, but natural forces start the repair process as soon as the next growing season begins. The eastern end, attached to Flying Point, has recovered from several blowouts in the last fifty years. One very large breach (punched through the central section of the dune in 1998) took only five years before sufficient sediment built it back up to half the pre-storm height.

MARTINIQUE

Flotsam mixed with dried sea grasses become anchors for new plant colonies. As they decay, the nutrients are recycled into the new growth.

When the waves trimmed the grass from the upper beach and dune ramp, they did not damage the deep roots. With that blowout, the flats behind the dune received an influx of new sediment, grass cuttings and other decaying material. The fan of overwashed sand actually created the preferred conditions for the clipped marram grass to sprout new shoots from the taproots. The storm even advanced the nest-building opportunities for the shorebirds.

The number of storms arriving on this beach, and the extent of the damage wrought by each, is remarkable. There are few beaches in Nova Scotia with a more aggressive storm-wave environment. Continuing pressure from storm events and increasing visitor demand have encouraged government planners to bring Martinique under their control in 1971.

Over the past several generations, the beach has maintained its tenuous attachment to the headlands and rocky ledges, even though storm surges and wind-driven waves have caused repeated erosion of the beach and dune. On the other hand, it was created by strong ocean currents and reinforced with the sand that was scoured from other locations. As with all beaches, Martinique needs to be in constant flux or it will succumb to invasions by the plant world, get covered over, retreat inland and disappear.

LAWRENCETOWN

Eastern Shore • 30 km east of Halifax

LAWRENCETOWN OFFERS ALMOST two kilometres of clean, hard sand and is only forty minutes away from Nova Scotia's major urban centre. In spite of that, this can be a lonely beach, even on a hot summer day. Beach lovers know that ocean breezes often lower the temperature at Lawrencetown to downright cool. Most sun worshippers tuck a sweater into their beach bag, in case a local fog bank wanders in over their picnic.

Visitors have also discovered that the waters of the North Atlantic are often warmer in late summer and fall than during the height of the vacation season. For that reason, Lawrencetown can often boast a September rush not experienced on other beaches.

Surf enthusiasts, however, know that "L'town" is great at any time of the year. They have rebranded Lawrencetown as the new mecca where the best rides develop after offshore winds blow from the northwest. Most avid surfers monitor a website operated exclusively for this beach and wait for word on the weather conditions that create the ideal swell angle. The waves tend to break both left and right on the reefs and sandbars, and surfers take careful measure of the rips and rocks.

Donning wetsuits and tethered to carbon fibre boards, athletes of all ages hit this beach when the tail end of Caribbean hurricanes wander close to

RIGHT: *When waves come straight in from well offshore (called a long fetch), they often draw sand back out to sea, leaving a long sloping ramp that dissipates the wave energy.*

ABOVE: *When the wind is off the shore, it confronts the waves surging in from the open ocean, delaying the break and extending the ride.*

RIGHT: *Aggressive storm waves driven in by passing hurricanes carry cobblestones well up the beach, overwashing the dune and boardwalks.*

OPPOSITE PAGE: *Waves on the southeast end of the beach can be head-high because of the depth of the bottom contours. They can become barrel shaped.*

the Eastern Shore. In the days before and after such storms pass, the North Atlantic sends terrific swells onto Lawrencetown Beach and the resulting waves can rival the best on the continent. International competitions are held here each year, like the Turkey Dip and the September Storm Surf Classic. Members of Canada's national surfing team regularly use the beach for rigorous training. The provincial Lifesaving Society is even hiring international surf guards to patrol the property.

LAWRENCETOWN

Lawrencetown sand is firm and tightly packed, partly because of the fine silt and sand mixture eroded from the coastal headlands like Half Island Point.

At high tide, the waves land with a crash upon the sand and stones, then retreat with a scrabbling sound that is unique to this beach. It makes this one of the noisiest beaches anywhere. The grassy headland up the highway just before the site is a favourite spot to assess the conditions. Surfers hope they hear that scrabbling sound as they perch on the knoll, surveying the beach. If they do, the surf's up!

The waters off Lawrencetown Beach are quite shallow, which would usually take the energy out of the ocean waves, but the storm surge here can be extremely aggressive. Ocean swells force the surface layers of the sea to move. Closer to shore, the waves slow down and squash together. This can increase the height of each wave and create huge plumes of spray. As the surf approaches shallow water, the bottom of the wave is forced to slow down as it feels the bottom. The top of the wave overtakes it, and it breaks.

Mounds of cobblestone are predominant at the eastern end and the back of the beach. It is also the underlayer along the whole shore. For most of the summer, the beach has a topcoat of fine, dusty sand over the stone, but in the fall, the shoreline is scoured by the frequent Atlantic storms. At those times, the sand from the beach is drawn back to submerged dunes. That tends to level out the seabed, making the surface waves more uniform too. The fine sand gets returned to the upper beach the next summer.

Offshore reefs, the ever-changing seabed and aggressive currents combine to produce undertows at the western end of the beach, called Lawrencetown

The picturesque headland is a familiar landmark and lookout for visitors to Lawrencetown. Its restaurant and sport shops are equally well known.

Point. It has also become known for rogue waves, which happen when faster waves from behind collide with slower ones, creating double waves and danger for unsuspecting surfers. This beach, and its surf, is always changing.

Whatever the conditions, August afternoons became so busy that concern for the safety of drivers, due to cars parked on the highway behind the beach, prompted the park management to create a visitor control plan. That included more services, like an expanded parking lot, freshwater showers, accessible boardwalks and changing rooms, which encouraged even more surf seekers! The beach is a top destination for surfers, on summer afternoons and year round. It is on the best of those days that the lifeguards are pressed to open two stations on the long sandy shore.

It is also a very non-commercial beach. Caribbean and European tourist destinations are known for rented lounge chairs, cold beer delivery and T-shirt sellers. In Nova Scotia, almost everything on a beach is regulated. If you don't bring it, it just isn't going to be there. That sounds harsh at first, but the cliffs rising on both ends of Lawrencetown shout out their unspoiled natural features while osprey soar overhead and dive into the surf for passing mackerel. There are no gated resorts, and "No Trespassing" signs are extremely rare. Some private homes have been built close to this shore, but the beach is free. It is free of almost everything except what nature washes ashore . . . and surfers!

RAINBOW HAVEN

Eastern Shore • 20 km east of Halifax

ON WARM SUMMER AFTERNOONS, the grey beach of Rainbow Haven becomes a magnet for local sun seekers. Even on misty mornings, at low tide, beachcombers come to wade in the quiet surf, sift through the sand and find sand dollars and other fascinating bits of nature's castoffs.

Reaching out from Pensey Head, the beach at Rainbow Haven appears to be a sand spit gradually creeping eastward. It became a barrier bar many years ago, almost completely closing off the Cole Harbour estuary. The spit is growing that way because of accumulating sediments, not just from the nearby cliff face, but from all along the coast, picked up and delivered by strong currents offshore. High tides and fall storms combine to create aggressive forces that undercut the exposed bluffs. The storms separate the sand and gravel from the underlying slate left millions of years before, and wash the fine, dark sand onto the beach.

Clumps of samphire greens and marram grass grow profusely at the high-tide mark. Drenched in salty fog and sprayed by the surf, they absorb the calcium, a nutrient they need to grow.

Adding to the organic mix of vegetables on the dune above are various species of beach pea, seashore buttercup and morning glory blossoms all colouring the landscape. Wild rose bushes are plentifully scattered in the lea of the stunted spruce trees, their succulent hips outlasting the harsh fall storms to provide a protein-rich meal for winter residents.

RAINBOW HAVEN

These dunes have a lush cover of plants and grasses. Their thick root systems stabilize the shifting sands, while the succulent leaves provide food and shelter for many levels of the food chain.

These plants are the builders of the barrier bar, offering stability for the newly storm-tossed sand, feed for a huge variety of local animals and shelter for the young of many insect species — too many insects for the pleasure of those sun seekers!

Of equal interest to the regular beachcombing crowd is the salt marsh created behind the sandy beach. Sheltered there and protected from the harsh North Atlantic storms is the Cole Harbour Marsh.

The marsh maintains a vital connection with the open sea via a surprisingly swift current running through a breach in the beach. Twice daily, the flow comes rushing into the shallow marsh. At full flow, the current must be treated as dangerous for young

In the waters just beyond low tide, a delicious feast of small plants and animals is continually brought to the surface by the waves.

swimmers, but this breach in the sand is extremely important to the growth of the marsh and all its residents.

As the cool ocean flows flood into the rich intertidal zone, they blend with the fresh waters coming down from the surrounding hills. Algae from the sea, mineral sediments washed down from the surrounding farmland and hundreds of species of marine life are exchanged with each tide. Twice daily, a full load of microscopic life is absorbed into the marsh, which recharges the food chain. It is a "vegetable soup" so nutritious that this marshland is one of the most productive ecosystems and fish nurseries on the entire coast.

Also flushing in with the tide are schools of gaspereau and smelt, eager to spawn and have their offspring nursed by the bounty provided by the warm, brackish waters. The minnow-sized mummichogs are particularly eager to locate the gooey rafts of mosquito larvae, their favourite food. Even more impatient are hundreds of shorebirds: herons, yellowlegs and sandpipers carefully inspecting the mud

RAINBOW HAVEN

ABOVE: *Many shorebirds nest just above the high-tide line. The Ipswich sparrow is hard to spot among the dense grasses and marsh greens. So much pressure is being placed on Sable Island, traditional breeding ground for the Ipswich, that more of them can be expected to extend their nesting choices and join the Savannah sparrow on the narrow coastal strips of beaches and salt marshes of mainland Nova Scotia.*

LEFT: *The seaside buttercup has a delicate flower of five tiny petals that blossoms throughout the summer. Many flowers, and even seashells, are divided into five, a common number in nature.*

near the low-tide mark for their favourite prey.

Burrowed just at the low-tide mark, periwinkles, clams, tiny shrimp and a variety of sea snails establish themselves in the sand and mud, waiting for their next meal to be washed over them by the inflowing seawater. With such a variety of life filling the marsh, it's hard to decipher which will be

ABOVE: *At the height of summer, even with its warm water and long stretches of beautiful sand, the park is never crowded.*

LEFT: *The showy morning glory is a hardy vine that thrives along the margins of a salt marsh. Its triangular leaves and pink, five-sided flowers brighten even a cloudy day.*

predator and which will become prey.

Terns, Savannah and even Ipswich sparrows nest in the grasses above the tide mark where their young find plentiful insects to feed upon, and protection from circling hawks and other birds of prey. The Ipswich sparrow is revered in Maritime birding circles, mainly because its primary breeding ground is on the fabled Sable Island. Many other migrating

species, arriving on their way to northern breeding grounds, find the marsh habitats an inviting place to rest and feed for several hours or days. Together, they make the marsh a wonderfully noisy space.

Even in winter, or perhaps especially during those cold, dark times, the marsh is a vital feature of life along the shore. When the dry land is frozen and covered with snow, the tide keeps the tidal flats open and relatively warm. There, on a chilly February morning, a tasty clam might be found and wrestled open, a fresh sprig of green sea grass might be nibbled on or maybe even a tiny shrimp-like creature scratched out of the mud, meeting the requirement for at least one square meal per day.

This beach, together with its tidal marsh, is always in transition. The rhythm of the daily tides moves sediments in from offshore, exchanges nutrients from the open ocean to the inner marsh and replenishes the garden it has become. Seasons mark the beach and marsh as well, at times offering up aggressive winter winds, then sweet summer breezes. This variety of changing environments is what makes Rainbow Haven such a sanctuary, a place where nature deals with the elements, survives most of the tests and recycles itself into the next generation.

5

CRYSTAL CRESCENT

South Shore • 30 km west of Halifax

PENNANT POINT EXTENDS MORE than five kilometres out into the open Atlantic, as the southernmost headland of Halifax Harbour. The whole peninsula has a tundra-type habitat and has resisted colonization by all but a few hardy woodland plants. That means the bare granite boulders are exposed, offering great hiking and a clear panorama of scenic ocean vistas. The irregular shoreline of this rugged coast features a naturalized thirteen-kilometre trail, numerous small coves and the historic fishing village of Sambro.

Often referred to as a coastal barren, the site could be described as pretty sparse. Scraped by glaciers and worn down by thousands of years of runoff, the limited soil barely covers the rock. It offers only enough depth for the hardy crowberry, clubmoss and dogwood to find root. Much of the soil and eroded pieces of granite are picked up and simply washed to the shore. Small pebbles get tumbled along in rough stream beds, their corners rounded off to create fine sand that then gets carried out to sea.

Offshore, the seabed remains shallow for several nautical miles and the waters are liberally dotted with ledges, shoals and numerous small islands. The sand eroded from the granite boulders on shore has

RIGHT: *The quality of the sand on a beach is determined by the rocks from which the material came. The same type of rock can even produce different colours, depending on the minerals within. Some grains are large and "sugary," some are a fine silt.*

ABOVE: *It's the fine dry sand that goes Aeolian. It usually seeks the upper beach, where it can build into a dune or overwhelm a walkway built to maintain a clear path.*

RIGHT: *The rhythmic washing of the beach by the waves creates crescent-shaped cusps of fine sand along the length of the beach.*

been settling in the shoal waters off Pennant Point since the retreat of the most recent ice cap, perhaps ten thousand years ago.

Shallow waters, shoals and small islands have a marked influence on the shoreline. They force the ocean currents to swirl around in the bay, mixing cold offshore seas and warm river runoff. This is where the deep ocean swells start to "feel" the bottom, picking up sediments from the seabed and delivering them back to the beaches, like in the three quiet coves of Crystal Crescent.

These three beaches face east, and while that is still

CRYSTAL CRESCENT

These beaches are the result of migrating sand being trapped in a sheltered cove, building up and covering the underlying rocks over time.

pretty much directly out to sea, the islands and reefs protect them. As the worst of the Atlantic gales pass by, eddies and counter-currents are broken off the main currents. The surging water slows down in the shallow depth, and its energy is reduced. This surf can no longer carry the newly scoured material from the bottom, leaving it onshore and covering the underlying rocks, adding another layer to the beach.

Many sandy shores depend on having a high dune ridge at the back of the beach to act as a storage locker for beach repairs needed after gale-force winds blow hard from the sea and carry away the sand. On Crystal Crescent, what would have been a dune is a low, narrow wedge of sand with its steep slope facing the sea, backed by huge, granite boulders.

Granite rocks are mostly made up of quartz, with glittery crystals of mica and feldspar. Feldspar is one of the most common elements on earth, sporting different colours in different places. This local variety sports a white tint, giving the sand on these beaches a clean, light colour from the water's edge right back to the rocky berm of the parking lot.

The coming and going of sand here does not seem to be adversely effected by the lack of a dune, however. Sand is regularly lost from all three beaches, particularly during the fall and winter storms, but the supply offshore appears sufficient to replenish the beach cover each summer season.

CRYSTAL CRESCENT

Sandy beaches are formed when sediment is accumulated, either by rivers washing it down from the cliffs above, by tides and waves that carry sand in from offshore bars or by currents carrying the sand along the coast until it is dropped in a quiet cove.

In fact, it may be the constant exchange of sand that keeps Crystal Crescent so clean. Other than the scattered clumps of dried eelgrass, there is little debris of any kind — hardly enough to sustain a population of hungry gulls. The cold temperature of the blue-green water usually horrifies visitors until September, when the tropical storms brush the Nova Scotia coast, but sunbathers rave about the beach and keep coming back to the park because the sand is so attractive, white and clean.

Washing up and down a beach, the sand also gets separated by grain size. The rhythmic waves create a softer layer with finer sand for walking and spreading a picnic blanket.

As long as the waves keep striking the beach at Crystal Crescent at the same angle, the fascinating pattern will repeat itself along the shoreline. As the arriving wave hits the horn of a beach cusp, it is divided, flowing in two directions. The friction slows its velocity, causing coarser sediment to be deposited on the edges. The waves then slide into the valleys and pick up finer sand grains. As they meet another wave in the middle, they flow back out to sea until met by incoming waves. The process repeats itself, wave after wave, and the beach takes on one of nature's classic geometric patterns.

Nature is even more apparent on the third of the three beaches at Crystal Crescent. Known locally as a naturist beach, it is at the end of the formal one-kilometre walking trail. Tall grass separates the trail from the beach and most people take advantage of the clothing-free option.

QUEENSLAND

South Shore • 50 km west of Halifax

QUEENSLAND WAS NOVA SCOTIA'S first fun beach. For many years, it was the place to be in the summer, and even considered "Ipanema North." Like the famous beach in Rio, the main roadway skirted the top fringe of sand, the trendy culture was young and hip and girls who could wore *hilo dental*, bikinis that were so tiny they seemed to be held in place by dental floss.

Just a thirty-minute drive from Halifax, and certainly the warmest beach anywhere within party distance, Queensland almost emptied the city on sunny days in July. The only problem was finding a parking place if you arrived too late. Anybody who could afford a cottage in nearby Hubbards just moved out there and walked the back roads to this beach.

There were two other beaches in St. Margarets Bay, but Cleveland Beach was thought to be more cobblestone than sand, and Bayswater was almost twice as far.

It measures only five hundred metres long, but Queensland attracted sixteen thousand sun worshippers over its short summer season. To assist with the crowds, the government purchased several beachfront acreages, allowing the roadway, shoulder and parking area to be widened by taking up what used to be the natural back beach. That just encouraged more commuters to make it their top choice for a summer

RIGHT: *The edges of the receding wave meet the next wave coming in. When the two forces meet, the friction causes multiple mini-breaks in the incoming crest.*

The popular beach at Queensland can almost disappear when the tide is high, particularly in the fall and winter, when the sand has been withdrawn back onto the reefs offshore. Summer waves will return the sand, however, and the popular beach will once again be covered by beachgoers.

afternoon. As they did, more people claimed, "There are too many people there, so nobody goes anymore."

Still, it offered some of the warmest water anywhere, very little fog and far less sea breeze than other Atlantic beaches. That meant warm summer afternoons for the family, sparkling sand and comfortably warm swimming. Being well up at the head of St. Margarets Bay and a long way from the open ocean, there was also less surf, which was a relief for parents of small children.

Like at other well-known beaches, the Nova Scotia Lifeguard Service has actively patrolled this strand for many years. Other efforts were made to accommodate the daily masses, including change rooms, picnic sites and outhouses. The strip of sand was never very wide, even at low tide and before the road was widened, so active sports like volleyball and frisbee were discouraged, reserving as much space as possible for the sun seekers. Parking continued to be an enduring concern, however.

In summer, the circulating ocean currents carry extra sand in from offshore and the beach seems wider and flatter. Incoming waves approach the beach at an angle, then drain back straight into deeper water, forming a zigzag movement along the shoreline. This is a naturally occurring pattern on many beaches, called longshore drift. When wave after wave follows the same pattern, they reinforce the regularly spaced depressions, creating cusps in the sand.

As the tide comes up on the beach, it also rises under the sand, making the surface layers much more fluid and unstable. The withdrawing tide removes some of that loosened sand as it departs.

With the sand transported away by longshore drift and redeposited on neighbouring beaches, more of the cobble foundation of the beach and even the road has been threatened several times.

Major storm events have become a concern too. During the big fall storms of 2003 and 2007, large, cobble-sized rock and even boulders were tossed onto the road and into the parking lot. Rows of armour stone were soon installed along the beach and a fresh coat of pavement applied on the main road.

At the time, the logic of rebuilding the road in the same place was questioned. Based on past experience at other Atlantic beaches, and factoring in the

During the fall storm season, big winds can push an incoming tide farther up the shore. The wave action then undermines the large boulders by dragging more sand back out to deeper water than the incoming waves can replace.

potential for increased damage from rising sea levels, it could have been the right time to propose a "managed retreat." The road and parking lot, it was suggested, could have been relocated, allowing the shoreline to migrate inland naturally. A back beach and even a dune ridge could have been allowed to form, making room for a larger or at least a more stable beach in the future, one that would naturally withstand the fury of future ocean swells.

In the lifelong struggle to master high tides and storm surge, beaches have been identified as the only really successful design. They absorb almost all of the wave energy that reaches them and do so in the shortest period of time. They may then be stable

the level of the beach is actually submerging as more sand is being carried away than brought back in.

Not far up the shore, Cleveland Beach stretches out across a small cove in the bay. It suffers the same winter and summer sand cycle. The gravel bar is probably lengthening, however, gradually accumulating more material in summer than it loses in winter.

The annual stripping and redepositing of material on Queensland Beach, plus the longshore drift there, will eventually result in a diminished beach. Over time, Cleveland's cobble beach berm could increase, benefitting from the reduction at Queensland, or just from the circulation and mixing of currents carrying sand around the entire bay.

If the waters circulate completely around St. Margarets Bay, the disappearing sand at Queensland could very well end up on Bayswater Beach at the bay's southern tip. Its sand cover appears relatively stable, at least more so than either Queensland or Cleveland, even though Bayswater is much more exposed to the open ocean.

Recent storm experience has underscored the lesson that an underdeveloped beach with wild vegetation on its dunes and a management plan that allows for not only storm damage, but naturalized repair processes, fairs better in the long term and costs far less to maintain. So the residents of Halifax may have to drive the extra distance to Bayswater Beach in the future. As expensive engineering fails to protect Queensland, and Bayswater is left to deal with the natural processes on its own, it may potentially be the new fun beach for the next generation of young trendsetters.

for decades. On the other hand, the next extreme storm could bring new change. Creating more room for beaches to change with the effects of extreme weather would perhaps have been a reasonable alternative to the expensive roadway repairs.

Sedimentary records over many thousands of years indicate that the sea level here has been rising slowly for a very long time. Only a few incidental records were kept until recent years, but it is also possible that

7

HIRTLES

South Shore • 120 km west of Halifax

THE KINGSBURG PENINSULA HAS BEEN A welcome sight for returning fishermen and other mariners throughout Nova Scotia's history. Bearing to port would lead them home to the seafaring communities of Bridgewater, LaHave and Riverport. A turn to starboard would mean Lunenburg, landing their catch and greeting old friends.

Today, the peninsula is host to new generations of homesteaders with modern lifestyles and a new appreciation for the weathered, sea-bound coast. It is also considered the poster child for a "living" shore, representing all beaches where coastal processes are taking a noticeable toll.

When the glaciers scraped across these shores, they deposited tons of gravel, sand and mineral debris. All this they held frozen until the most recent era of glacial melt. Kingsburg's headlands, like Point Enrage, contain the piles of material that they left behind.

Hirtles Beach, the crescent-shaped cove in the lee of Point Enrage, faces the open Atlantic full on. The slope of the beach is quite gradual, which does take some of the energy out of the incoming surf, but there is nothing else to protect this beach. The ocean is wide open from the low-tide mark all the way to Spain.

The long beach is now a sand and cobble bar with a few dune ridges at the seaweed line. It had its start as a small spit in front of two coastal lagoons. One of them, Romkey Pond, is still connected to the ocean by a small stream, and depending on the

The Kingsburg Peninsula faces the Atlantic Ocean full on. Recent hurricanes like Juan and post-tropical storm Noel have each made changes to the layout of Hirtles Beach.

amount of fresh water it receives from its neighbour, Hirtles Pond, it may be more salty one day than the next. Both ponds are usually fresh water, but surging waves can overwash the low dunes and turn them very salty.

Protruding out into the open Atlantic as it does, Kingsburg Peninsula and these beach ponds often attract passing waterfowl. As many as fifty-seven species have been recorded by avid birdwatchers. The frequent switching from brine to fresh water and back again has reduced the volume and variety of plant life in the ponds, however — a fact not favoured by the hungry birds.

The dune is made up of gravel and coarse sand, which allows much of the incoming water to gently seep through the porous mixture, reducing the amount of sand lost to erosion. If the dune were tightly packed with fine sediments, it would better resist the initial tidal surges, but high-energy waves would still, eventually, wear it away.

Despite the fact that full-blown hurricanes and post-tropical storms have all made changes to the structure of Hirtles Beach, it and the dunes are in good condition. Each big wind scours the sand, reduces the dune width and height and even pushes it in towards the land up to a metre, but it persists.

The storms that harass the beach also erode the headlands, and their coarse glacial till is the material supply needed to rebuild the beach and dune after each storm passes. Historic records of Hirtles Beach support the theory that there is sufficient sand offshore to rebuild the beach slope after each storm, which it seems to accomplish naturally. Unfortunately, it takes a few years longer to rebuild the dunes and backshore. That explains why the beach appears stable, but the dune ridges are

TOP: *Great Blue Herons are common along Nova Scotia's coastlines. They particularly enjoy the variety of salt and fresh water fishes available in the many salt marshes and tidal pools.*

BOTTOM: *Wild storms of the last decade have scoured sand from the dune, reducing its width and height and resulting in an eleven-metre march of the beach landward.*

relatively smaller than they were a generation ago.

Sometimes a healthy beach like Hirtles seems like a desert with very little life. It receives continuous supplies of silt and sand from the headlands and shallows nearshore, however, and teeming amounts of microscopic marine life. Nothing stays still on this beach, though. The next wave either covers it up or takes it away.

There is no best time to take in the splendid scenery of Hirtles Beach. Even during major weather events, the sea comes alive with energy and the surf delivers classic breakers. Between storms, the ocean water is crystal clear. Rays from the rising sun shine through the cresting waves that roll in on this dynamic beach.

Owned by the local government, Hirtles Beach offers extensive parking, wonderful trails and swimming, all maintained by an active community committee with the goal of preserving it for any visitor who appreciates the site.

If there is ever a favourite time in nature, it is the hours just before and after sunrise. Most beach inhabitants seek their main meal at those times and venture out of their dens before the sun is too high. The most common sights on Hirtles Beach are the footprints of contented visitors, both human and animal.

RISSERS & CRESCENT

South Shore • 130 km west of Halifax

SAND HAS BEEN ACCUMULATING ON Rissers Beach since well before European explorers first recorded their visits.

In July 1607, Samuel de Champlain sailed into what we now call Green Bay. He and the crew were quite taken with the beauty of the site and stayed four days, carefully noting the natural features of the shallow water anchorage and the possibilities of establishing a new colony.

Green Bay is indeed a lovely and peaceful haven, but the quiet waters belie the constant circulation of the sand. With the causeway at the northern end, much of the sand and sediment just circles the bay, building up on Rissers Beach and growing a barrier bar there that almost closes off the mouth of the small river recorded by Champlain as Petite Riviere.

Taken as a whole, Rissers Beach, plus the growing sandbar at Petit Riviere and the crescent-shaped causeway to the north, form a very popular beach system. Much of it is a provincial park with a busy campground, picnic area and a very enjoyable stretch of sparkling silver sand. Summer temperatures on Nova Scotia's South Shore can be fairly warm too, so many vacationers choose to spend their holidays here at the provincial park.

This peaceful shore saw tremendous upheaval, however, millions of years in the past. Softer sedimentary rocks, like the layers of sandstone that formed eons ago, were tilted and pushed to the surface. The glacier that covered the rocks for thousands

Green Bay has been attracting visitors for more than four hundred years. It continues to be a popular stopping-off spot for shorebirds, families of summer campers and all manner of marine life.

of years weighed the rock down then scoured the surface and opened fissures between the softer layers as it retreated.

Today, daily tides and annual freeze–thaw cycles grind down these mini-crags and small pinnacles, leaving crevasses where sand gets trapped and sea creatures can hide.

Brown and green seaweeds are robust enough to anchor their durable holdfasts into the fractures of these rocky outcrops on the lower beach; even the most energetic waves rarely dislodge them. By absorbing the energy from the crashing waves, the rocks and sea plants protect the upper beach, where a maturing forest of hardy species maintains a tenuous hold.

ABOVE: *Yellowlegs are common visitors to the back beaches and marshes in spring and fall. They fish endlessly for small minnows and aquatic insects, which keep them well fed in preparation for the winter. Where yellowlegs spend their time in the summer breeding season is less well documented.*

RIGHT: *The conflict between incoming sea and the encroaching forest growth is played out at the high-water mark.*

The gnarled trees will eventually lose their battle with the shifting sands, as tidal surges undermine even their thick root mass, but not before new healthy trees can establish themselves nearby and attract deposits of additional sand.

While they stand, the mature softwoods separate the sandy shoreline from a small marsh behind. To explore both, visitors are encouraged to follow the long curving boardwalk. It steps over possible nest sites and a variety of sensitive plants while leading into the waterfowl feeding areas.

When occasional Atlantic swells do approach the upper bay, most run afoul of several shoals and

Seaweeds come in all colours. When leaves of fragile dulse fronds are ripped from their beds just offshore, they tinge the backlit waves a blood-red hue.

sandbars. At those times, healthy colonies of more colourful seaweed are cropped by the aggressive surf. If backlit by a rising sun, the red algae adds an attractive hue to the breakers as they reach the beach. Once on the shore, the bits of dulse, sea lettuce and kelp moulder and rot, enriching the sand. Tiny beach fleas and other small shrimp-like creatures gather in and under the piles of wrack, enjoying the safety of the moist debris. In the evening, they venture out onto the bare sand, boldly defying the shorebirds that seek out the tasty prey.

Crescent Beach, the causeway across the top of Green Bay, is actually a two-kilometre-long, gently curving sandbar connecting the mainland with the LaHave Islands. It is a true tombolo, a naturally occurring spit of sediments that gradually reaches out from one point of land to another over time. These structures form in the lee of islands, where wave action is reduced. Tombolos are very temporary, though, or at best, very unstable.

The dunes of this structure are badly eroded, with its high point, covered in abundant marram grass, remaining only at the southern end. The whole system would have failed long ago but for the paved road running behind the dunes and the wooden retaining wall along the seaward side. Over the last century, the

BEACHES OF NOVA SCOTIA

ABOVE: *What the sea has created, it often attacks. Continuing efforts have been put into shoring up the causeway since 1905, including loading in armour stone, trees and even old car bodies.*

OPPOSITE PAGE: *In what seems like an impossible location to survive, seaweeds of the intertidal zone seem to thrive. They take the incoming breakers in their stride and actually sap some of the energy from the waves, protecting the shoreline.*

man-made barrier has been reinforced with car bodies, rocks, trees and a makeshift seawall.

Not being a designated beach and outside the park boundary, Crescent Beach is the only beach in the province where cars are permitted to drive and park. Actually, that is a great advantage to some clam diggers and seaweed harvesters. To curb the somewhat destructive uses of motor vehicles, local residents on the islands have formed a community committee that tries to manage and protect the stability of their land bridge to the mainland. The destructive force of even a single ocean storm or the inexorable rise of sea levels will surely bring changes to the entire beach system in Green Bay. The committee cannot stop those changes, just hope to manage them.

9 CARTERS

South Shore • 160 km west of Halifax

CARTERS BEACH IS OFTEN DESCRIBED as the most beautiful beach in Nova Scotia, and it may be! It is truly the undeveloped shoreline of fable.

It is really three small, crescent-shaped beaches, each joined to another. The first two are separated by a thin, deep brook, called a runnel because its course across the shore wanders from side to side as sand is washed onto the beach and into its path.

All three crescents are blessed with powdery white sand and the best chance to find sand dollar shells anywhere. Gnarly black spruce trees maintain a foothold in the sandy dune ridges adorned by a variety of beach grasses and unusual lichens. What looks like paradise really does have a lot to recommend it, but the beautiful green water is most often quite chilly, and when it is warm, jellyfish are likely nearby.

Carters is unspoiled and much like it has been since the most recent ice age receded. The few signs of human establishments in the area are actually from centuries of aboriginal encampment.

Provincial park planners have decided to keep it completely undeveloped, and they cite several good reasons, including a rich cultural history, ecological sensitivity and its relatively small size. There is a published plan to give it a "Nature Reserve" designation, the highest level of provincial protection

RIGHT: *The solid granite that underlies this part of the province has contributed to the fine, white sand.*

CARTERS

available. The special status would extend to neighbouring Wobamkek Beach, as well as three islands: Spectacle, Jackies and Massacre.

Local residents are concerned that even if no development and none of the standard services are made available, not even an increase in the size of the small parking lot, visitors will still come. That will create pressures on the privately held lands surrounding the beach property and the shoreline environment will suffer. Officials counter by saying that ecological protection is a delicate balance and that the limited "carrying

ABOVE: *What looks like dried up weeds is an important mix of sea grasses and lichens that bind the sand together and provide nesting materials for shorebirds.*

LEFT: *Each spring, a deep runnel is formed on the beach as fresh sand being washed ashore interferes with the runoff from ponds behind the dunes.*

capacity" of Carters Beach must be safeguarded.

In the end, private citizens, the local community and every visitor will be responsible for enjoying the area and ensuring that it is left in its natural state. Wise public stewardship of Carters Beach will bring its own rewards.

10 SAND HILLS

South Shore • 100 km southwest of Liverpool

THERE ARE AT LEAST FOUR CANADIAN provinces with a "Sand Hills" park and even a variety of Canadian wines with the same name. This park, well down the sou'west shore in Nova Scotia, is the only one that can offer an actual beach to go with the sand hills.

Spectacular dunes give this park its name, although most of the really high ones are well back into the treeline. Still, some of them are ten metres high and offer lovely, out-of-the-wind sites for a wine and cheese picnic.

The presence of these huge dunes, a good walk from the current beachfront, indicates that the shore is not where it used to be several hundred years ago. Today's beach is a more recent phenomenon.

Construction of the causeway to nearby Cape Sable Island likely caused a shift in tidal currents and longshore drift in Barrington Bay, resulting in the rapid growth of this shore and its iconic sand spit. The older dune ridges at the back of the new beach continue to be eroded, but the windward side of the beach receives ample new sand with each incoming tide, and so the dune endures.

Abundant pioneer plants, like the sandwort, trap the sand and bind it with their roots. The decaying humus they create becomes new soil that traps rainwater. Less hardy plants now move in. As more plants colonize the dunes, the sand is covered and the dunes change from yellow to grey. Now strengthened against incoming tidal surges and blowing sand, the

ABOVE: *Even though the older dune ridges of the back beach continue to be eroded, the current beach receives ample new sand with each new tide.*

LEFT: *Pioneer plants trap sand, binding it with their roots. The humus they create adds to the fertile growing conditions, and the new soil traps rainwater so less hardy plants now grow.*

back beach, with its mosses and surface lichens providing a more solid "crust," has a grey appearance. This is the natural aging process for a beach.

Long ripples and the gentle slope here are indications that this beach is accumulating sand from other areas. Just how gentle the slope is can be seen at low tide, which seems to go out forever! When it comes in, the ocean waters are warmed by the sand and especially enjoyed by toddlers. It's a good thing the beach is continually being replenished here, for it always seems like buckets of sand are taken away at the end of the day — in the kids' bathing suits and car floor mats!

BEACHES OF NOVA SCOTIA

SAND HILLS

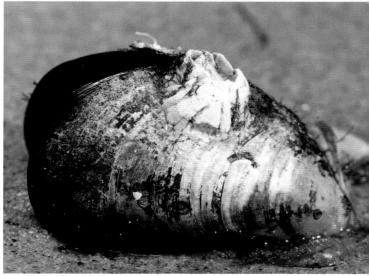

ABOVE: *This barnacle has permanently cemented its relationship with the mussel. The two animals feed on basically the same things, so as long as the mussel remains submerged, both will survive.*

LEFT: *Changes to the shoreline down at Cape Sable Island may have caused a shift in currents and resulted in the growth of a new sand spit at Sand Hills.*

This is a favourite destination for hundreds of families in the summer. It is also one of many Nova Scotia shorelines that hosts thousands of sandpipers during their autumn migration, plus a few pairs of piping plovers during their nesting season. This poses challenges for planners when designing the visitor services for the beach. Families have come to expect vault toilets, cook shelters and raised boardwalks, but park planners don't want to create "hidey holes" — places that could accommodate foxes and other predators that might stalk the plovers.

Whether for fish, fowl or family, Sand Hills may be a long way off the main road, but it is worth every minute of extra travelling!

THE HAWK & CAPE SABLE

South Shore • 80 km south of Yarmouth

RENOWNED AS THE SOUTHERNMOST point of Atlantic Canada, and for the very productive fishing grounds nearby, The Hawk on Cape Sable Island is little more than one long sand-dune ridge sitting atop solid bedrock that is over five hundred million years old! The beach canvass is a colourful result of the eroded sandstone, shale and the hardened greywacke that underlies the region brightened up by tiny fragments of seashells that dot the tableau.

Numerous hummocky dunes grace its eastern shoreline, offering some profile and forming a series of barriers connecting small islands with salt ponds and marshes behind them. The whole of the long, sandy point appears stable, but the shoreline is being continually redistributed by winds and the prevailing currents, which circulate sand from beach to beach. It may even travel as far away as Sand Hills Provincial Park at the top of Barrington Bay.

The pronounced hummocks attest to the frequent sea breezes that buffet Cape Sable Island, moving the sand and reshaping its beaches. Other than these dunes, the whole of Cape Sable Island really has no high point, or any place to escape the blowing sand. It really is a lovely "island of sand"!

On the eastern coast of this island, South Side

RIGHT: *Sand covers almost everything. Beneath the surface — somewhere down below — are the same foundation rocks as elsewhere across Nova Scotia: greywacke, sandstone and slate.*

BEACHES OF NOVA SCOTIA

THE HAWK & CAPE SABLE

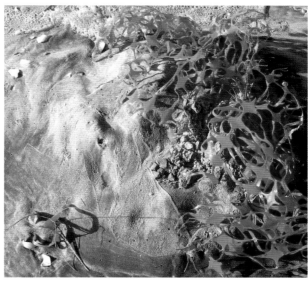

ABOVE & LEFT: *South Side Beach is that classic deserted strip of sand known only to local residents, who protect their hidden gem at the far end of a rutted track with no signage.*

and its beach once supported Canada's highest dunes at almost thirty-six metres, but historic sheep grazing reduced grass cover and the dune system eroded. Today, South Side Beach has very few sheep and sees almost no visitors. It is hidden at the end of a rutted access trail, and the people who journey here endure the rough approach knowing that this abandoned beach is a stretch of paradise. Because the beach is next to the fish plant, industrial litter occasionally fetches up here. Neighbours accept that as a part of living with a vital commercial fishery, however, and continuously collect the debris as part of their environmental stewardship.

Beautiful as they are, the beaches of Cape Sable lie completely exposed to Atlantic storms. They rarely have a protective blanket of snow or ice. In fact, they may not even have been covered during

The geological record shows that Cape Sable has been below sea level a number of times and also well above the high tide mark for extended periods. A sunken forest of well preserved roots attests to long-term cycles in sea level rise and fall.

the most recent ice age. Yet they survive the worst of the winter's icy gales. Scientists who have studied this look to the ripples in the sand. They can tell from the patterns if enough sediment is moving in to keep the features of the beach actively cycling. Ripples are a sign of a healthy beach and The Hawk and South Side Beach are in good condition.

The critical bird breeding areas the beaches provide are less appreciated but as important as the rich fishing grounds. Most bird species that appear at The Hawk do so in the thousands. In fact, it is one of Canada's designated Important Bird Areas for the diversity of shorebird species that use it as a short-term stopover along the bird migration route known as the Atlantic Flyway. A series of very large and impressive interpretative panels on the beach explain why Cape Sable Island is a major staging area for species on their seasonal journey. It is also the site where small numbers overwinter.

One of the reasons they come in such numbers is the low dune ridges, which are frequently over-washed by storm surges. That allows for a good inflow of nutrients to the marshlands behind. The eelgrass growing on the mud and salt flats is the attraction for the avian invasion — at least for the Atlantic brant, a smaller species of goose that stops here to fatten up on its way north. Protecting these feeding grounds for the brant and so many other species is crucial to their survival.

If there were ever a place to hide in plain sight, it is the beaches here — particularly for the piping plover. In fact, several pairs return to nest on the open sand here

Shorebirds, in their multitudes, converge on Nova Scotia beaches as part of their long migration south. These short billed dowitchers are fairly common at this time of year as they use the beach as a fly-in restaurant.

each year, along with countless other breeding birds.

Much is being made of sea-level rise and the fact that parts of Nova Scotia's coast appear to be sinking due to movements of the Earth's crust. The geological record shows that Cape Sable has been below sea level on a number of different occasions, and even well above the high-tide mark for extended periods. A sunken forest of well-preserved roots and stumps attests to long-term cycles in sea-level rise and fall. Exposed tree stumps in an area called the Drowned Forest, said to be fifteen hundred years old, are imbedded in a peaty soil, offering tangible examples of how shorelines continue to be adjusted by overwhelming geological forces.

When these trees were living, Cape Sable Island was part of the mainland and the low-water mark extended offshore for a full nautical mile. Not stable enough to maintain that connection now, the current sandbar that stands as an introduction to the "Island of Sand" is little more than a wandering spit, never able to anchor a permanent causeway from the island to the mainland.

Storm days bring lobster fishermen to this beach, not in search of lobster, but to recover their gear — driven ashore by offshore gales. These beaches receive constant winds and several major storms every year. The Hawk and South Side Beach are surviving those torments, to the unending appreciation of a few beachgoers, the many migrating birds and the beaches themselves.

MAVILLETTE

Bay of Fundy • 40 km north of Yarmouth

MOST OF THE SHALLOW BAYS AND inlets along the shores of St. Mary's Bay have protective offshore bars that have been building up and wearing away for millennia. While many act as important sand storage for local beaches, experts in coastal management agree that these bars, and even the beaches they feed, are continually moving. The fear is that most of the sand is migrating towards the shore. As the sand moves, the beach seems to disappear and the salt marsh behind it fills in. Shorefront properties are at great risk. It is possible that very few of the beaches we know today will endure these gradual pressures, let alone the encroaching human activity of the next few centuries.

In order to monitor this thesis, numerous researchers have studied beaches and their attached dune ridges along the entire Atlantic Coast. Most of their measurements indicate that beaches do move, from natural causes, and that property owners should have a real concern about the shifting sands. The shoreline at Mavillette, however, with some of the highest and widest dune ridges, seems to be unique. It has been relatively stable for a long time.

Large dunes, over five metres high and two hundred metres wide, dominate the fifteen hundred metre stretch of fine, grey, sandy beach. Early mapping shows that Mavillette was originally formed as a spit of sediment creeping out from the eroding cliff face to the south. That sand spit now forms a

Cape St. Mary's is more than a headland; it's a solid promontory jutting out into the bay, trapping the shifting sands and protecting the beach and its marsh from the high energy of Fundy waters.

This five-metre-high dune ridge was established many years ago and has withstood the fury of countless Fundy gales.

ABOVE: *It's hard to tell at first if a hard-shell clam is alive or dead. A gull will likely pick it up and drop it on a roadway from a great height to see if there is a meal inside.*

RIGHT: *Long established boardwalks protect the dune vegetation, giving the yarrow and hardy rose bushes a chance to colonize the area with their stabilizing root systems.*

solid barrier almost completely across Cape Cove, the lagoon in the shadow of Cape St. Mary's.

Adding to its mass are the huge tidal flats. When the water is low, walkers can wander more than six hundred metres out to sea, and barely get their feet wet. This feature isn't lost on local residents who appreciate the way the cold Bay of Fundy waters are warmed up as they flood back in over the exposed sand. Clam diggers also take advantage of the distant low tides, seeking the hard-shelled clams as a garnish for their traditional rappée pie.

Anchoring this barrier beach are several ridges of stable dunes, thickly colonized by wild rose bushes and even yarrow. If they are moving, it appears as if they are accumulating more sand, filling in the breaches and actually moving slightly farther out into the sea.

What used to be a salt marsh behind the beach is not filled with sand, but with compacted black soils

MAVILLETTE

and layers of well-aged vegetation thick enough to allow only a small creek to drain the low land.

Early settlers once harvested the salt hay that grows between and behind the dunes, destabilizing the high ridges and giving ocean gales a chance to create blowouts in the dunes. Sand was mined here for a few generations too, and even more tragic was the one-time passion for dune buggy racing, which made tracks across the land that were still visible on the dunes until recently.

The area has been a provincial park for over two decades, and while thousands of visitors still

Sediment studies of the Mavillette marsh have exposed layers of peat and other plant materials. These layers encourage the forest growth now colonizing the edges, offering hundreds of birds, like this snipe, feed and cover.

congregate at Mavillette each summer, effective management controls have reduced the destructive activities. Wildlife and wildflower habitat has been enhanced as well, as more nesting sites are protected from the unnatural forces that shaped the landscape.

Ocean waves rarely overwash these dunes, assisting in the beach stabilization. Each year, storms still blow in on the full length of the beach. That gives a few eager surfers a chance to test their skills near the cliffs, but most of the shore area is very shallow and breakers that do rush up the beach are usually sapped of their energy, having dragged their feet for six hundred metres along the fine sandy bottom.

With very little ocean water rushing in, the thousand-acre marsh is no longer very salty, and while the land is still poorly drained, the natural succession from salt marsh to peat bog, and then to a more fine-grained loam, has definitely started to take over.

Plant species that now chose to move into the area are more suited to a damp forest environment than a swamp. A broad mix of tamarack, black spruce,

MAVILLETTE

ABOVE AND RIGHT: *Avid birdwatchers come to Mavillette to see herons, snipes and willet — migrating species that look for grass and woodland habitats characteristic of mature marshlands.*

and red maple is expanding, attracting upland bird species like snipe and willet during their migration. Wildflower meadows too are expanding, attracting different birds and insects. As the opportunities for observing the wildlife expand, so have the efforts of the modern park management. Protective boardwalks have been created throughout the wetland and interpretative panels in the bird observation tower help visitors get more from their trip to this beach than just a sunburn.

13

BLOMIDON

Bay of Fundy • 120 km north of Halifax

IT IS TOLD THAT, MANY YEARS BEFORE recorded history, an industrious beaver built such a huge dam that it closed off the upper Bay of Fundy between Cape Blomidon and Parrsboro. The Mi'kmaq man-god Glooscap worried that the pressure of water held back in the Minas Basin would become dangerous, so he broke the dam and the water flowed out with such force that it bent the tip of the North Mountain backwards into what is now Cape Split.

One positive result of all that water rushing back and forth past Cape Split and Blomidon has been the erosion of sand and mud, washing down from the cliff face onto a vast, red-sand beach. A great but gooey tidal expanse has grown into a fertile habitat for an amazing variety of sea creatures. It is a wonderful outdoor laboratory.

Considered to be the end of Annapolis Valley's North Mountain, Cape Blomidon was formed during volcanic times, well before the native Mi'kmaq settled the land. It has a cap of hard basalt rock, underneath which are layers of sandstone and shale. Nearer the base, where the tidal surges attack the rock, shallow sea caves appear close enough to be open for exploration.

As the softer sediments have been lifted and tilted,

RIGHT: *Some cliffs are 180 metres high, almost twice as tall as any building in eastern Canada. The exposed layers of sand or mud show the many intervals of sea-level rise and the tilting of the Earth's crust.*

The sediments that wash down from the clifftops to the sea floor contain plant and animal nutrients. That enriches the intertidal zone, where rockweed and barnacles find a home.

vertical cracks have developed across them. Added to that is the scouring of the last ice age, as recently as thirteen thousand years ago, and rising sea levels, which began again here four thousand years ago. Attacked by four millennia of tidal action, the soft layers of the cliff face have continued to fracture and erode, undermining the cape even further. Yet Blomidon endures.

Twice daily, the Fundy tides expose colonies of rockweed and barnacles in the shallow tidal pools. They cling to the few solid rocks that have tumbled down in the waterfall from the upper basalt layer of the cliffs. The leafy seaweed provides cover for myriad microscopic creatures that need to remain wet during low tide. High tide then returns to flush them out and, if they are unlucky, into the swaying tentacles of the waiting barnacles.

A closer examination of the sand at low tide reveals holes and pock marks where worms and tiny copepods have burrowed to stay wet. These little mud shrimp feed on even smaller bits of plant and animal materials, which have come down with the erosion of the cliffs above. This turmoil of the sediments and the tides spread out over the broad fertile beach creates the perfect mix of ingredients for a highly productive marine ecosystem.

This beach and the neighbouring Evangeline Beach are not just about what lives in the sea, however. They support over a hundred species of birds, including thousands of sandpipers and plovers at a

ABOVE: *Thousands of sandpipers and plovers play follow the leader, first wheeling to the left and showing their underbellies, then to the right as their backs come into view.*

RIGHT: *Semipalmated sandpipers rest between furious eating and the next synchronized flight.*

single low tide. Their sole reason for being there is to peck out the teeming numbers of mud shrimp that don't burrow deeply enough into the exposed seabed to escape their long, sharp beaks. Several plover and sandpiper species depend on these wriggling, rice-grain-sized creatures in late July and August, when they must double the weight of their tiny bird bodies in order to endure a four-day flight to wintering grounds in South America.

Some August days at Blomidon and its neighbour Evangeline Beach are absolutely overwhelmed by a celebration of shorebirds. They flow back and forth over the vast flats like spiralling ribbons precisely controlled by a rhythmic gymnast, or get herded

Eagles, like their osprey and peregrine falcon cousins, were once uncommon visitors to these shores. Now they preside over an uneasy hierarchy of year-round residents.

back towards the cliffs, roosting more and more densely as the tide rises. At other times, even at the height of their migration, there seems to be scarcely a creature; perhaps, scared away to a neighbouring beach by a peregrine falcon surveying the shore from a nearby perch.

The height of the tides in the Bay of Fundy is well known. What visitors sometimes underestimate is the speed of the incoming water, which can flood ten metres of the beach flats here in as little as a minute! Fortunately, Blomidon is more sand than it is mud, and beachcombers can scurry away from the rising tide without getting stuck in deep ooze. Evangeline is more muck than sand.

Now part of a popular provincial park, Blomidon offers a number of interpretative programs and receives more public attention than other parts of the Bay of Fundy. Volunteers often lead visitors in

LEFT: *The escarpment is receding quite rapidly here, but only partly due to the tidal action notching out the bottom layers. Small rivers gathering storm water often flow fast enough to gouge pebbles and larger-sized rocks from the cliff face.*

ABOVE: *The shorebirds are drawn to the tide's edge where the mud shrimp are plentiful. Birdwatchers are drawn there to photograph the action.*

activities like "Birds of Prey in Nova Scotia" and "Muddy Blomidon Biodiversity," interpreting the natural history of the site. Informative panels describing the local wildlife are effectively located at the top of sturdy staircases that stretch from the parking lot to the beach. The wonders of the Bay of Fundy ecosystem are all on display at these two beaches.

Alongside the living attractions on this shore, and actually littering the beach, are semi-precious stones like amethyst, jasper, agate and gypsum. These fall to the base of the cliffs and attract legions of rockhounds. Often collected from the cliffs and beaches outside the park boundaries, they can become multicoloured ornaments and jewellery. The Mi'kmaq people say that they were gifts from Glooscap.

BLUE BEACH

Bay of Fundy • 20 km north of Windsor

BLUE BEACH, NEAR HANTSPORT, IS not a beach as most visitors would picture one. It is an outdoor museum with much more to offer visitors than soft sand and gentle surf.

The shoreline is called a shingle bar, and it attracts international fossil enthusiasts because it is covered — actually littered — with traces of prehistoric life. Experts in the field say the impressions and traces found here represent a significant collection of trackways perhaps made by the very first four-footed animals to ever walk on dry land.

Fossil discoveries were first recorded here in 1841. Since then, significant finds are made regularly, even by youngsters in school classes who have simply turned over rocks to find the imprints of jawbones, teeth and amphibian skeletons. Seasonal outflows from the upper Avon River also wash layers of gypsum and limestone from similar deposits inland, adding a greater variety of fossilized marine life, even coral.

When the great continents separated, sections of the earth's crust began grinding together and layering. At one point, Nova Scotia was submerged like an undersea bowl where silt collected. As the global temperature rose and fell, shale, siltstone and harder rocks fused together under great pressure, creating patterns and fracture lines.

In the intervening ages, the great oceans dried up, leaving marine life stranded, which then fossilized in the sediments. When the planet cooled,

If you stand quietly at the foot of these cliffs, you can hear a constant shower of mini-landslides. It's the layers of shale and other sedimentary rocks, which contain the fossil record of early life here.

The many sediment layers at Blue Beach are tilted and lifted into view. The fossil contents seem to fall out, having suffered little damage over the millennia.

BEACHES OF NOVA SCOTIA

Blue Beach has the characteristic signs of a former salt marsh, the bottom of a frozen lake, the edge of an ancient sea and even a fertile field. All these epochs of the planet's history come alive here on the exposed shore.

ice advanced from the pole, weighing the northern half of the continent down, effectively creating the first of many epochs of sea-level rise. Then, when it gradually melted away, the earth's crust rebounded, creating dry lakebeds and more layers of sediments.

Through these tumultuous periods of the continent's history, the different layers hardened, were tilted and eventually thrust upward into view. The siltstone and shale cliffs now exposed at Blue Beach represent the lower layers of those basins, originally laid down about 350 million years ago. This site was likely the shoreline of a tropical salt

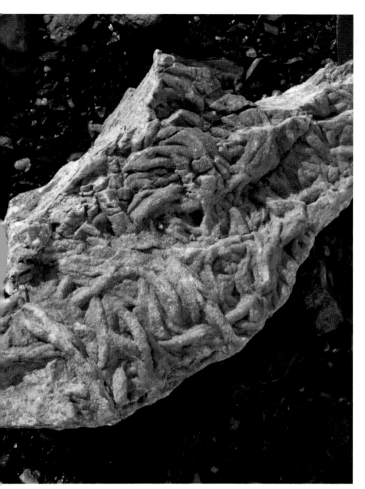 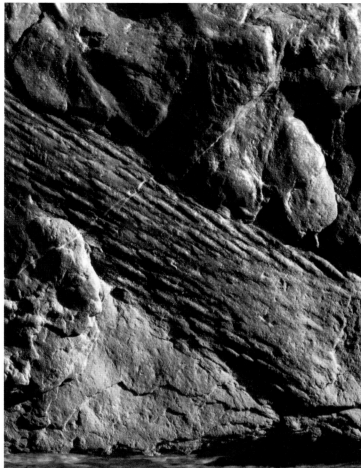

LEFT: *Many animals without feet also called the beach their home, like the worms that made these burrows three hundred million years ago. They are ancestors of species that live there today.*

RIGHT: *Plant fossils, particularly ones from large trees like the* lepidodendron, *which covered Nova Scotia in the Carboniferous Period, are also common finds.*

marsh, teaming with plants as well as animals, plus the sediments that recorded their movements.

It's not that Blue Beach was the first or only place these animals could have walked ashore. It just happens to be a site where their arrival is very obvious. The beach here runs in the same direction as a major fault line. It has weakened the bond between soft and harder rocks. Exposed by waterfalls and washed by the tides, the cliffs of loose sediment generate showers of fossilized animal, plant and raindrop imprints.

Each time a small cascade of shale tumbles down the cliff face to the beach, new discoveries are made. With each incoming wave, shards of rock are flipped over, often exposing animal parts or imprints

ABOVE: *Many fossils are from the early Carboniferous Period and represent the most significant collection of four-footed animals recorded anywhere.*

LEFT: *The bluish-grey shale and clay of the cliffs erode and become the cover layer of Blue Beach, giving it its name.*

OPPOSITE PAGE: *Long rows of rocks look like dinosaur spines radiating across the beach. In reality they are tilted layers of sandstone, but they too contain fossils of long extinct creatures.*

— many the size of small coins. It is now child's play to sift through the deep layers of shale and uncover traces of life many millions of years old.

The march of time measured for each geological process seems very slow, sometimes hundreds of millions of years, but it took only an instant for the plants, fish and tiny mammal-like reptiles on Blue Beach to be trapped and become a future fossil.

BURNTCOAT HEAD

Bay of Fundy • 100 km north of Halifax

THERE ARE TWO BEACHES HERE THAT are easily accessible and very dramatic. Don't expect to spread a blanket on a stretch of white sand at either, though.

The sight of rushing tides and swirling waves is part of what brings visitors to Burntcoat Head. The spectacle of millions of tons of seawater flooding into the Bay of Fundy every twelve hours or so is unique in the world.

The other part of what makes Burntcoat Head so fascinating is what can be seen on the beach when the tide is out.

When all the churning and swirling ocean water has moved on, an expanse of dry land stretches for leagues beyond the shore, dotted by countless red hummocks and beds of soft, green sea grass. The hummocks are hardened deposits of sandstone that resist the strong tides. The sea grass is an algae called zostera. It is secure in its place on the sandy surface, holding this part of the seafloor from washing away. It too is highly resistant to the harsh winds and cold temperatures, which are often a part of Bay of Fundy weather.

At Burntcoat Head, the beach is not a beach; it's the floor of the sea, between the high- and low-tide lines. It is acres of mudflats and innumerable shallow pools hosting an extremely fertile ecosystem. The seafloor teems with marine life even at low tide. A tasty buffet of a million small creatures is kept moist in the topmost layers of mud, or under

It looks like slippery mud but it is actually red sand, offering firm, dry footing for eager beachcombers.

a cover of sea grass, waiting for the ocean to return and flood their home once again.

Huge flocks of shorebirds descend upon this cornucopia as a priority stopover on their long annual migrations. It will most likely be the sandpipers, frantically racing back and forth, that gain the largest share of exposed seafood. The hungry hunters could also be a flock of herring gulls or a family of raccoons teaching their young that a low tide means dinnertime. The exposed seabed is particularly important for wildlife during the winter. When all of the onshore land lies covered by ice and deep snow, the beach may be the only place to find food.

It is easy to spot the hard shells of mussels and crabs, but the bulk of the beach residents are much smaller. Worms, tiny mud shrimp and microscopic plankton are there in number, filling the shallow pools or burrowing in the mud. Some animals prefer the upper beach, where the water is warmer, but they must remain exposed for longer times. Farther out on the seabed, the mix of species changes. Colourful starfish, pudgy anemones, tiny fish and shrimp seek out the lower zones, where the soft mud and rock ledges are the first to be covered again by the incoming tide.

Where else can you stand upon the bottom of the sea, stroll amid the barnacles and lift huge fronds of kelp to see what is hiding underneath? Habitats like

At this point, Bay of Fundy tides can cover as much as a metre of beach in ten seconds

these are living observatories, offering easy access and opportunities for identifying live sea creatures only seen in schoolbooks.

The ocean rises and floods these shores like clockwork. The earth's rotation and the moon's gravity maintain the daily cycle, like pushing a child on a backyard swing. The extreme narrowing of the Bay of Fundy and the Minas Basin along with the very gradual slope of the bottom squeezes the water high up the cliff face, and it is that vast horizontal change that can leave a wide vertical expanse of the ocean floor exposed.

Much of the Bay of Fundy shoreline is red sandstone, created when many layers of sand and silt

The greatest vertical rise and fall of tides ever recorded on this planet exceeded 16.5 metres, and it occurred right here at Burntcoat Head!

hardened into sediment millions of years ago. The cliffs are made up of two enduring elements, feldspar and quartz, but the dramatic component is iron oxide. It tints the sand and keeps the waters a reddish-brown. It also leaves indelible stains on a beachcomber's clothing, an inescapable souvenir of a trip to Burntcoat Head.

Summer rain and winter ice wear down the top edges of those ancient sandstone cliffs, and the daily rush of the tides and waves erodes these bluffs from their foundation up.

There is a picturesque community-sponsored picnic park and interpretative centre in the old lighthouse building on the tip of Burntcoat Head. It beckons visitors to this shore. They hunt for shells, semi-precious stones such as agate and amethyst, and record shorebird sightings. Fossils too, relics from the Triassic Period of two hundred million years ago, can be found among the layers of sandstone.

By the way, the name of this promontory has nothing to do with a burnt coat. Like many place names in this region, it is a mispronunciation of the original Acadian name: Pointe Brull or Burnt Coast.

16 BLUE SEA

Northumberland Shore • 80 km west of Pictou

THIS PROVINCIAL PARK IS OFFICIALLY Blue Sea Beach, aptly named because the summer sun off Treen Bluff turns the water a lovely ocean blue. Sitting out near the end of Malagash Point, the five-hundred-acre site has a wide beach that makes that blue sea very accessible, plus a pleasant picnic area and at least a minimum of visitor amenities.

Separating the unspoiled beach from a huge wetland behind it is a single, low dune — high enough to maintain the level of water, but low enough to ensure that occasional storms will flood in from the Northumberland Strait with extra nutrients for the animals there. That wetland is Beatty Marsh, which gives this beach the name more commonly used by those who know it.

Whichever name it goes by, the beach is well known as a perfect place for summer fun, birdwatching and as a nursery for several species of small fish.

High tides, which occasionally overwhelm the beach and its low dune, nourish the salt ponds with seaweeds and creatures fresh from the ocean. At those times, mats of plankton, insects and juvenile fish get caught in the tidal pools, perhaps finding refuge until they can follow a return wave to the sea — or becoming a meal for a marsh resident. This is the first level of an important wetland ecosystem.

Several species of fish actually need a freshwater or just slightly salt environment to raise their young. The salt concentration is not a factor for the adult fish themselves, as they can change back and forth,

The low dune ridge here keeps the calm, brackish water of the marsh separate from the waves of the Northumberland Strait. This is important to the fish and shorebirds, and a pleasant alternative for small children looking for a natural wading pool.

but it is important to the specific food the young fish need at the early stages in their life cycle. Often, the important lower levels of the food chain, like bacteria and microscopic worms, are easiest to find in marshlands.

As an outdoor laboratory, tidal pools like MacInnis Pond (near the picnic area) have long been magnets for children. Filled with warm, shallow water, they make it easy to spot and capture mummichogs, silversides and thousands of tiny fish that dart back and forth chasing the nutritious plankton. For children, these wetlands provide endless summer hours of play and these future biologists never come home dirty, just soaking wet!

This picturesque shore is sheltered by the narrow Northumberland Strait and is protected by a solid layer of ice for several months. In such a low-energy environment, the beach maintains a very flat slope. Low dunes are often associated with long flat beaches because the surf has lost much of its force by the time it has reached the upper beach and dropped the sand it had picked up along the way.

Here, the incoming surf has a higher concentration of fine silt, which remains floating in the surf longer. As it nears the shore, that silt drops out too and creates rows of closely spaced ripples. Minute particles of sea plants, carried in on the waves as well, settle between the ripples to feed the clams and other shellfish that enjoy the shallows.

With the slowly advancing tide, the sun quickly

ABOVE: *Shellfish feed only when the mud above them is covered by at least a skim of water. Snorkling in the warm, shallow water makes it easy to watch for their telltale holes.*

RIGHT: *Most species of fish can live in either salt water or fresh. Their preference depends on the stage of their lives. Brackish pools next to a saltwater beach offer both.*

warms the water. Shellfish, like clams and quahogs feed as the incoming tide washes in a flood of plankton. Observing them, then, requires a bit of extra equipment, but waiting for their siphons to appear and watching them feed is a favourite pastime at Beatty Marsh.

The low, sloping beach can be a bit of a nuisance, particularly at low tide. For visitors wishing to launch a canoe or kayak rather than swim, it's a long carry to

Tiny worm-like creatures that thrive in the beach mud come to the top of their burrows when the tide comes in. They think they are protected by the skim of water, but the sharp eyes and long pointed bills of the sandpipers are more than a match for these wriggling morsels.

the deeper water. Once afloat, the sea, sand and scenery of Beatty Marsh shoreline becomes idyllic.

Even in calm settings, some parts of the beach get washed away, exposing the sandstone bedrock and the crevasses carved in its layers. On a falling tide, several of the salty pools are filled with tiny creatures, exposed and hoping to avoid the hungry gulls waiting for their chance at a fresh meal.

The combination of wetland and ocean shoreline has become vital for breeding fish, nesting birds and healthy wildlife habitats. The link between these is the beach where tidal flushing starts the process.

Highly productive ecosystems like Beatty Marsh host the plants that flourish in the brine and the nutrition that arrives with it. Few animals actually eat salt marsh plants, but after the plants die, they decompose in the warm water and are absorbed by the lowest levels of the food chain, bacteria and single-cell animals, and eventually the young fish. These then sustain the higher levels of the food chain, capped by the adult fish as they escape across the beach and back out to the blue sea.

17 RUSHTONS

Northumberland Shore • 40 km west of Pictou

THE TRIP FROM THE PARKING LOT TO THIS peaceful shoreline features a trail and a long boardwalk skirting an attractive wetland — not a wetland the way the word is commonly used to make a swamp sound more attractive, but a pleasant landscape of ponds, cattails and dune ridges. This is where the wafting aroma of ocean breezes and the strident squawk of a red-winged blackbird greet you.

With the terrain remaining quite soggy most of the summer, the boardwalk provides access across the sensitive bird-breeding habitat, and the observation deck becomes a stakeout to spy on the nesting sites.

All summer, the submerged root systems of the abundant water plants host hordes of tiny pond creatures. The seeds of the late summer bulrushes are a favourite morsel for young ducks, and muskrats seek out the tender parts of their root mass. Bulrushes grow in abundant clusters with their long, flat leaves providing excellent cover.

Sites like this beach park are well positioned along the Atlantic Flyway route to be key stopping points each spring, offering safe haven for waterfowl, shorebirds, upland species and even birds of prey.

Almost all of the Brule Shore waterfront was bought up several generations ago by city dwellers looking for inexpensive lots along a family-friendly seashore. Most cottage lots are still passed down in wills. Rushtons Beach Provincial Park is one of the few public access points along this shore, but few visitors put this area on their "must see" list. On

ABOVE: *An observation deck provides good sightlines to waterfowl nesting sites.*

LEFT: *The red-winged blackbird is well known in Nova Scotia's marshlands, seen here standing guard over an unseen nest.*

the other hand, visiting shorebirds like the spotted sandpiper love this site, seeming to know that they can get the protection they need.

Stepping off the boardwalk, onto the last low dune of warm sand, the beach seems forever calm. This is not a large strip of oceanfront, nor a particularly soft and sandy one. The fine, dark grains of sand are firmly packed. Protected by the Cape John headland, the sheltered environment of Rushtons Beach has very low-energy waves. It rarely receives

ABOVE: *This is not a large beach, nor a particularly sandy one. The fine grains feel more like silt than sand, densely packed by the waves.*

RIGHT: *These spotted sandpipers are fairly common, but there are many species that stop here, looking for just the right nesting sites.*

new sand from wind-driven waves or storm surges. Here the clear, curling waves move very little sand around in the swash zone. This shore seems asleep, hypnotized and immobilized in the gentle surf.

Ripple patterns are always interesting on mud and sand shores. When submerged by a few inches of incoming tidal water, and with the setting sun glancing off the surface, they can take on a ghostly aura, like the footprint of a passing astronaut or perhaps an ancient fossilized trilobite.

LEFT: *Rushtons beach is not well known, but it is one of the few nature preserves along this shore. It is cherished by both two- and four-footed locals.*

This beach has a very gentle slope, continuing well offshore. The occasional storm waves that do bring in seaweed and shellfish from deeper waters provide beachcombing fun, but the great family entertainment here is in the warm shallow water — splashing, wading and learning that it is easier to float in calm salt water than it is at home in the lake.

Behind the dune ridges, picnic shelters and fire pits are placed away from the shore to accommodate marshmallow and wiener roasts, the perfect way to end a long summer evening here.

18 MELMERBY

Northumberland Shore • 30 km east of Pictou

MELMERBY BEACH DOESN'T BELONG to the land or the sea. It is a part of both, being at times a seaside people-magnet, then a mere pebbled margin, with much of its fine sand temporarily lost to the onshore gales and a punishing storm surge. Such is the continuing life story of Pictou County's favourite beach.

While it shows evidence of having survived for thousands of years, Melmerby is forever growing and receding. It first stepped out into Little Harbour after the recent ice age. As a narrow spit of sand, it was fed by sediments delivered from Kings Head by the currents of the Northumberland Strait. Over many centuries, it joined with Roy Island, becoming a tombolo, a sandy barrier connecting that island to the mainland. All the while, the beach was evolving into high ridges of dunes. These were often cut by storm-driven waves, but also repaired naturally with the sand brought in by subsequent gales, and stabilized for periods long enough to acquire a sparse crop of marram grass.

The winds in the Northumberland Strait, together with the surface waves they create, bear down on the many sandstone bluffs that mark the coastline. Fine sediments are eroded from the cliffs, blown into the sea and carried along by the currents until another headland is reached. Then the waters are trapped in

RIGHT: *The predominant current moves westerly along this coast, refreshing the soft sandy beach with new materials from the headlands.*

a backwash that leaves the fine sand deposited on the shoreline. That is how Melmerby beach was created, in a hundred tempests — the result of strong currents and wild onshore breezes.

The winds and currents that carry the migrating sands to Melmerby, however, also take the sand from the beach to be redeposited in the next crescent-shaped cove. The actual grains of sand seen on this beach today are not the same ones that were there last year, nor will they be resident for long. Sand is always moving on. Without such a continuing longshore drift, Melmerby would cease to

MELMERBY

ABOVE: *In some places, fine silt gives the sand a more dense composition, marked by long channels carved by receding waves.*

LEFT: *Fencing keeps people from trampling on the dune grasses, and that slows the shifting sands, but nature continues to move the dunes around.*

be able to repair itself; its dune ridges, if breached by autumn gales, would never have the materials to refill the gaps.

Where the sand movement is most evident is in the swash zone. It is there that the white froth washes endlessly back and forth between the tide line and the sea. The bigger waves often carry sand and foam far up the beach, but as they recede, they drag most of it right back into deeper water. Only the powdery sand is left at the high-water line. There it can dry out and become airborne, perhaps ending up as part of the dune ridge. The larger grains are rolled

Actually, aggressive breaking waves are rare at this beach because the shore is so well protected. Still, these ankle-deep waves will bring new sand to nourish the beach and wash away dead plant materials to feed the creatures in deeper waters.

back into the swash zone where they are continuously being churned about. Unable to pack together well, the sand grains in the swash zone remain a soft slurry where a beachgoer's feet sink deep. This is the sand that migrates along the beach.

What forces the sand to move sideways along the beach, and not just back and forth like the foam, is the fact that almost all the waves actually rush onto the beach at an angle, but recede straight back into the sea. This zigzag pattern of the waves carries the sand and pebbles with it, in small steps, along the length of the shore. Creating a shoreline river of watery sand, the currents carry the sediments on Melmerby Beach like a pianist doing a *glissando* on a keyboard, slipping down the shore from one end to the other, and then from one beach to the next.

Melmerby Beach is fairly flat, and the near shore is shallow for a long way out, punctuated by several sandbars. When the sea is calm and no wind is present to move the water, small waves still appear, crisscrossing the beach, moving only the fine grains of sand. Here they can fill the gaps between the larger grains of sediment, actually forming a surface layer of clay on top of the sand. As evening approaches and the receding tide leaves tiny rivulets in the surface, Melmerby takes on the cloak of a summer

paradise, sand and sea in quiet harmony.

Beaches are almost always windy places, however. In summer, the air warmed by the daytime sun will cool quickly as it moves out over the ocean. Changing tides and fluctuations between day and nighttime temperatures give birth to the onshore winds, sometimes growing into stronger gales.

Even here, especially in the fall, an extreme gale can bring violence to this beach. During one wild October in 1890, the barque *Melmerby*, carrying fifteen crew and a full load of lumber bound for Greenock, Scotland, foundered here, sadly giving its name to the beach and a lesson to all in the power of the sea. As recently as the last decade, violent winds moved so much sand around the beach that some structures were buried and others completely displaced. These were but reminders of the natural factors that preside over our beaches. They are formed by tempests and are continually shaped by those same aggressive forces.

The one advantage Melmerby has in comparison to beaches along the Atlantic Coast of Nova Scotia is its winter ice cover. The Northumberland Strait is icebound from December to April, and while ice pans may scour the beach, no high winds can blow holes in the snow-covered dunes. No fierce gale or surging waves can rob the beach of its sand.

Melmerby has grown and survived for several thousand years. If the river of sand keeps flowing past its shores, it will remain thousands more.

POMQUET

Northumberland Shore • 20 km east of Antigonish

POMQUET BEACH CONFOUNDS THE experts because it is actually *accumulating* new sand due to the natural action of storm events, wind-driven waves and high tide surges. Dunn's Beach and Monks Head are part of the same system along the Antigonish shore, and while they are eroding, this beach is growing new dunes and stretching farther into the sea. Pomquet is one of a few such beaches, growing with each storm as sediment from nearby headlands settles upon the shore.

Almost three-and-a-half kilometres long and eight hundred metres wide, it has some dune ridges as high as a modern home. In fact, Pomquet is the largest single dune system on mainland Nova Scotia. The highest ridge is the newest and most seaward one, while most landward ridges are low and are sometimes inundated by tidal flooding. Several small salt marshes and barachois ponds have been created as a result.

Scuba divers say that these undulating dunes repeat their pattern offshore several more times, becoming shallow sandbars. Ocean currents carrying the sand and gravel from nearby Monks Head are forced to slow down as they run into this shallow water. They are sapped of their energy and drop a good deal of the sediment load.

Pomquet Beach Park is shaped like a Thanksgiving Day turkey drumstick. Thanks to the modern boardwalks here and interpretive panels, it has been transformed into an open-air, living classroom. Perhaps

POMQUET

When many beaches seem to be migrating towards the shore, this beach is actually accumulating new sand and adding row after row of dune ridges.

the most interesting panel explains that this beach is an example of "dune succession," a process of storm-created dunes building up from sediments washed onto the beach. Raising your eyes above the boardwalk, it is easy to see how successive dune rows show the transition from wet sand at the water's edge to a mixed forest on dry land.

Walking farther along the boardwalk, visitors cross the hills and valleys of successively older dunes, each showing more fully developed vegetation. Hardy marram grasses hold the sandy ridges together until other seaside plants can move in. While that is happening, the unpopular poison ivy and prickly wild rose bushes keeps visitors from disturbing them.

In a natural process usually blamed on sea-level rise, high winds and waves are moving most shorelines farther onshore. This has the effect of forcing the dunes of most Nova Scotia beaches to invade the treeline. In areas of adequate sediment supply, some dunes are able to retain their form, and in a

few cases such as Pomquet, dunes are being added to the system.

When high tides combine with gale-force onshore winds, that sand is washed up onto the shore, building yet another new dune row on the beach. Of course, wild wave action can also draw the new sand back into the sea, but that process is usually associated with winter storms, and at a time when the beach at Pomquet is covered by a thick blanket of protective sea ice and snow.

The overall increase in sand deposits has essentially created a huge barrier across Pomquet Harbour, attached to the mainland at the northwest end, all covered by an ever-growing dune forest. Despite a great deal of man-made disturbance in the past, including dune buggy races, sand removal

ABOVE: *Poison ivy doesn't hide its showy foliage — resembling many other leafy plants that are not nearly as troublesome.*

LEFT: *The landscape seems to open up from one side of the horizon to the other.*

and vast burnt areas, the vegetation is now healthy. Scientists say that this system is a recent formation, probably developing during the last thousand years. According to the oral history of the local Mi'kmaq community, however, this dune has been a sandy shore much longer than that.

Thirty-nine plant species have been identified on Pomquet dunes. This is not surprising, since the presence of multiple dune ridges creates a wide variety of habitats, from exposed and salty environments near the shoreline to sheltered areas between and behind the dunes. The variety of shrubs and vascular plants is much larger on Pomquet than on most beaches and includes a few species that are found nowhere else in the region.

Like several of Nova Scotia's shores, Pomquet Beach has recently been favoured by piping plovers. These shorebirds prefer to locate their nests on open

At the height of summer, the park, with its warm water and long stretches of beautiful sand, often remains uncrowded.

coastal beaches and shorelines, but because those same areas are frequented by human beach enthusiasts, the piping plover has a poor record of nesting success. This has resulted in a designation of endangered in Canada. The good news is that Pomquet regularly hosts up to twelve breeding pairs of these plovers, which is almost 3 per cent of the known Atlantic Canadian population of the species.

Many of Nova Scotia's other signature shorebirds are also common here during migration season. Some of the most recognizable include the gangly great blue heron, plus family groups of osprey and eagles.

Pomquet is a modern name for this beach, taken from its mixed Acadian and Mi'kmaq heritage. Long before the arrival of Europeans, Native residents referred to it as "Popumkek" or "Pogumkek," meaning a place with lots of sand and bushes. It was transcribed by the Acadian settlers into "Pomquette," a name they could comfortably pronounce, and everyone who comes here still knows Pomquet as a wide, sandy beach.

A few years ago, several of the local landowners, including descendants of the Acadian families who received the first land grants, cooperated to sell their beachfront properties to the provincial government.

That led to the establishment of Pomquet Beach Provincial Park and significantly enhanced protection for the plants and wildlife. Since the piping plover has become a favourite local resident, it was taken as a symbol of hope in the community and has been adopted as the beach mascot by students at a local school. The breeding areas are well marked during the height of the season, and vehicles are not allowed on the beach. Given the size of the nesting territory demanded by each breeding pair, however, it is unlikely this beach will be able to support many more than are currently hosted, so the number will likely not increase here in spite of careful protection.

Pomquet Beach Provincial Park is a gem. There is a gated access road leading to a large parking lot. Vault toilets and a change house are well placed, and signs of good maintenance are everywhere. With the acquisition of the historical land grants, the beach is now protected from cottage sprawl.

Still, this beach is an enigma. At the height of summer, with its warm water and long stretches of beautiful sand, it is rare that the park would be described as crowded. Nature is left undisturbed at Pomquet, unexposed to human interference. Maybe that is a good thing.

BAYFIELD

Northumberland Shore • 25 km west of Antigonish

THE SEA IS VERY KIND TO THE diminutive Bayfield Beach. Open to St. George's Bay, and a long reach offered to both wind and wave from the Gulf of St. Lawrence, this beach could receive aggressive storm surges and flooding tides, but the waters here run shallow, far out to sea.

A long, sloping shore diminishes the crashing waves that would cause heavy beach erosion. As well, the tides on this shore are small, averaging less than the height of a large dog, meaning even when they are high, the grey sandy beach handles them with ease.

Like Rushtons and several other beaches along this shore, the waves, now sapped of their energy, can carry only the lighter grains of sand and silt, creating a beachscape that is a blend of clay and fine grit. These materials bond more firmly than would coarse sand and make for firm footing, even on the lower beach. They also make for the most curious ripples and patterns left by receding gulf tide.

At first glance, the stretch of sand looks like a desert, but the damp shore is teeming with living creatures: thousands of tiny clams, worms and shrimp-like bugs that burrow into the beach to avoid predators as soon as the tide has dropped. That is true for most beach dwellers, but a few remain up

RIGHT: *The footprints of time are left by the outgoing tide, only to be replaced by new works of seaweed art when the next wave rises.*

ABOVE: *The upper beach is often littered with natural debris, cast there by wind-driven waves or a rare tidal surge.*

RIGHT: *The tide takes most of the sea creatures out with it, but a few remain. The common periwinkle leaves tracks in the sand, made at a speedy one metre per hour.*

in plain view. The common periwinkle, with a hard, round shell it counts on for defense, leaves an obvious track in the sand. As it moves, it filters the top film of watery silt and captures unseen morsels of food. Children, if they can stand still long enough, get a great kick out of watching the mollusk come out of its shell and slide its way slowly along the ridges of sand.

In the mid-beach zone, the sand grains are larger and mixed with small pebbles, rounded stones and flotsam and jetsam. All this has been tossed up on the higher part of the beach by occasional big waves capable of carrying larger objects. The sand is less firmly packed here; it is softer and warmer, offering a better place for a picnic blanket, games of frisbee or fetch with the dog.

Lamb's quarter thrives in the debris of the upper beach, where it finds some solid ground to withstand the tide and wind.

Larger, cobble-sized rocks are scattered on the upper beach, remnants of a long-ago storm. The really big waves sometimes have the force to carry the large materials up that far, but rarely have the energy to drag them back down to the lower beach.

Clumps of seaweed suffer the same fate. Ripped from their beds in the shallow depths nearshore, they are carried high onto the beach and left at the strand line when the storm subsides. Dried in the sun and colonized by hordes of sand fleas, these clumps of beach debris attract flocks of hungry birds. Their nearby nests and the clutches of eggs within feed hungry families of raccoons, whose droppings in turn feed other tiny creatures that live buried in the sand…the cycle of life on a beach.

Over time, all this activity actually stabilizes the

shoreline, creating more solid ground, less likely to ever be washed away. The hardy lamb's quarter is an example of that, planting firm roots deep into the beach above the strand line, creating a new colony and hoping to trap other materials tossed in around it. The more variety, the better when trying to build a new base.

Beach communities and their territory are never secure from the onslaught of nature. They are forever being relocated, resculpted, joined and severed, but never diminished. They are like mats of seaweed, arranged as footprints and left by the outgoing tide, only to be replaced by new works of beach art when the next wave washes in.

PORT HOOD

Cape Breton • 50 km north of the Canso Causeway

AS EARLY AS THE 1780s, PORT HOOD was a productive little settlement with a burgeoning fishing, mining and agricultural economy. The large island that protects its busy harbour was the source of much of the fine quarried stone used to craft the windowsills and door frames at Fortress Louisbourg. That island was recorded then as attached to the mainland, but by the mid-1800s, the natural causeway had been severed in order to encourage the shipping of goods. Continuing erosion has widened the gap in the passage.

In 1958, an attempt was made to reconnect the island to the mainland, but little sand or other natural sediment ever accumulated around replacement boulders and it remains separate. In spite of the lack of sand for rebuilding the causeway to the island, Port Hood has five sandy beaches — although a few have gravel and cobble mixed in with the finer sand.

The attractive beaches on both sides of Shipping Point must have received most of the sediment scoured from the former causeway. It is a triangle-shaped dune that juts right out into the harbour, and the only one in Nova Scotia with such a shape. Nowhere else has such a dune just appeared and held its ground, apparently not swinging either way.

The exposed edges of the dune are more or less stabilized by plant growth. The soils are still quite poor and storm surges do undercut them, but the total structure presents a nice backdrop for a sunny picnic site on either side, depending on the wind direction.

The sandy shorelines of Shipping Point can be warm and soft on one side and a bit stony on the other, depending on the prevailing breezes.

Shipping Point, or Boardwalk Beach as residents call it, can brag that it is bathed by the warmest ocean waters anywhere in Canada. Lifeguards regularly record the water at 22–24 degrees Celsius. Benefitting from the protection of the headlands and its large island, Port Hood has a temperate climate, summer and winter. At low tide, visitors walk out to the sandbars or dig for clams. If the wind is blowing and the temperatures are a little cool, it is a great spot for kite flying.

More than a dozen species of dolphins, pilot whales and porpoises have been listed as regular visitors near Cape Breton beaches, pursuing schools of small fish. Cruising all over the North Atlantic, they are usually spotted when they take a breath or two before diving for two minutes at a time. Travelling in gams of up to a dozen, young males are driven away from the group at an early age to form their own family group. It is not uncommon for them, or older animals, to appear on the shore here, having died of natural causes.

Adding to the unique character of this beach, it is backed by a series of solid dune ridges and a "dune slack," which is, again, rare in the province. That term refers to a large, low area behind the dune that has gradually been colonized by hardy species that bind the sand together with their roots. As they grow and trap rainwater, more species move in, using the

ABOVE AND RIGHT: *At Port Hood Station, the mature dune slacks support healthy cranberry beds. They produce a bountiful crop of fruit each fall for the local residents.*

organic material created by decaying plants.

The most popular pioneering plant is the bog cranberry. It thrives on the combination of thick sphagnum moss, acidic waters and peat, all covered by a spongy, mat-like surface. It has spread profusely over the bog's surface, much to the delight of local harvesters.

Pathways across dune systems are always a threat to their stability, but here, the dune-building plants that have colonized the habitat include hardy sedges and heath like sheep laurel and cranberry. They give

A substantial six-hundred metre boardwalk has been constructed to prevent erosion of the dune and to direct beachgoers to its very tip. Several pathways lead off the track, however, allowing pickers access to the valuable cranberry bog.

the dune stability and discourage new trailblazing.

In an effort to protect the dune ridge, and to offer easier access the cranberry bog, the boardwalk has been constructed, starting very near the parking lot and leading to the far end of the point. There was also an attempt to stabilize the dune with the Japanese or beach rose. It was originally thought that since that species of wild rose is particularly resistant to dune blowouts and tolerant of adverse environmental conditions, it would be a useful addition. Now considered an invasive and even noxious weed in coastal areas adjacent to sand dune systems, efforts are being made to remove it before it displaces native rose bushes.

Although popular with local residents, very few actual tourists ever seem to take the time to stop over at this spot. Funny, when did golden sandy beaches with abundant nature, a boardwalk and warm seawater lose their appeal?

WEST MABOU

Cape Breton • 60 km north of the Canso Causeway

THE BEACH AT WEST MABOU IS LESS than a thousand metres long and the dune behind it is nearly as wide. This suggests that the beach has been stable and in this location for a long time. More fully described as a barrier beach and dune system, it even appears to be growing larger, with new sand ridges climbing up to the cliffs behind it.

These high dunes hold back a ten-kilometre-long harbour and salt marsh behind the West Mabou Beach. Home to a great variety of waterfowl, marine mammals and marsh plants, the area is a vital link between the land and the sea.

As with most of our beaches, a defiant spit first steps out to cross an open inlet. Growing into a sandbar, and stabilized by an emergent crop of marram grasses, this one almost made it. As it accumulated deposits over thousands of years, the barrier beach (and an offshore sandbar of poorly sorted gravel that has grown to join it) would have completely closed the estuary but for the heavy spring runoff from the West Mabou Highlands. This keeps the long, narrow harbour open to the sea.

The shape of the beach shows how it has been weathered by repeated offshore winds. It has a gradual slope towards both the land and the ocean side and has not been undercut by aggressive waves, which would have notched the part of the dune facing the sea, uprooting the plant cover and carrying the sand to the offshore sandbar.

High-energy waves that erode other sandy shores

WEST MABOU

Sea foam, or spume, is created by churning seawater. It contains lots of tiny plant and animal debris, which gives it its colour. As the breaking waves arrive at the surf zone, they trap air, forming bubbles that stick to each other.

ABOVE: *Usually looking for tender shoots and safe grazing sites for their young, deer often roam this Cape Breton beach at night.*

RIGHT: *This barrier beach and dune system appears to be growing into a sizable sandbar stabilized by an emergent crop of marram grass.*

are not a factor here. West Mabou is exposed to stiff breezes but Prince Edward Island is quite close, so the strong winds do not have a chance to generate destructive breakers.

In summer, the beach enjoys warm offshore winds rather than sea breezes. This brings fine sand back down from the gently sloped dune, covering the larger pebbles and creating the smooth, evenly sifted beach enjoyed by summer visitors. Gentle currents near the shore and endless waves also carry fine sand in from the offshore sandbar. On a sunny Cape

WEST MABOU

Like at Bayfield and Rushtons beaches farther into the Northumberland Strait, winter ice and snow cover these shores and afford the protection needed from icy winds and storms, which could rob the beach of its sand — and there is little enough of that near the southern end.

Breton day, it is one of the most preferred locations for a family picnic or teen beach party anywhere.

Winter storms could very likely attack these dunes, but a full blanket of snow and ice protects them from the worst of the weather. On the other hand, small ice pans that remain late into the spring regularly scour the beach, carting quantities of sand out to sea as they drift away.

Along this coastline of western Cape Breton, big winds often blow in from the Gulf of St. Lawrence, but the bedrock is resistant to erosion, providing little new material to feed the beaches. Frost does open up small cracks in the rocky highland terrain where the rain seeps in, but the minerals can be quite hard and take a long time before they are carried down to the sediment-starved beach.

More of the local sand seems to be generated from the crumbling cliff of glacial deposits around the harbour. Debris, rock and rubble at the bottom of the beach are a combination of many softer materials, which likely originated in the several hills surrounding Mabou. That variety creates sand grains of different sizes and shapes. With different-sized grains, a range of cavities between the particles is formed, from microscopic to large enough to accommodate a small creature. A great many marine species can find

WEST MABOU

Common sights on Nova Scotia beaches, the smaller herring gull and the great black-back are often seen together, but easy to tell apart. The immature of both species are virtually identical.

a personal niche on beaches like these. Tiny plants, and the almost invisible animals that feed on them, find the sand of West Mabou beach a perfect blend of habitats, not to mention the shorebirds that compete for a spot in the food chain.

As the breaking waves curl onto the surf zone, they trap air, forming bubbles that stick to each other through surface tension. The surf picks up the myriad of tiny plant and animal debris from the beach, colouring the foam and making it appear frothy. Sea foam, or spume, is frequently created by churning seawater where there is a natural loading of organic remnants.

The harbour channel, held open by the natural runoff as well as a large breakwater, sees a constant flush of this organic material exiting the salt marsh and washing up on the beach. It adds to the food chain along the nearshore, attracting an expanded list of marine species and creating a living laboratory of ocean ecology. The local beach management committee and the provincial government have cooperated to install interpretative panels at the parking lot and a nature trail leading off into the back beach and marsh. All that enhances the beachgoing experience for the local residents and the visitors fortunate enough to discover West Mabou Beach.

23

INVERNESS

Cape Breton • 80 km north of Canso Causeway

INVERNESS BEACH STARTED AS A sandbar across the mouth of MacIssac's Pond. Over countless centuries, it was supplied by ample quantities of sandstone and silt particles carried down to the shoreline by melting snow from the nearby highlands. Also fetching up here are the fine sediments carried in on the ocean currents that circulate around the southern Gulf of St. Lawrence. As a result, Inverness has become a natural repository and offers a sugary blond sandscape northward for almost four kilometres.

Summer temperatures are quite comfortable in Cape Breton, but no visitor has ever complained of oppressive heat on these beaches. The good thing is that the water temperature is often warmer that the air. Inverness Beach, like several others in the southern Gulf, offers the warmest salt water north of the Carolinas. It is a magnet for local families, who can't wait for balmy days to spread out the family picnic when the sun is high and the winds are off the shore. It is also the perfect setting for young couples to enjoy warm summer evenings and sunsets, which have become legendary here. When the summer winds do barrel in from the ocean, active sports enthusiasts know Inverness is perfect for kite surfing.

RIGHT: *Summer breezes are much kinder than the les suête winds. August's warm waters and sugary sands draw grateful visitors to this shore.*

INVERNESS

ABOVE: *Winds are almost always blowing on Inverness Beach. With the short summer season, visitors enjoy the beach in as many ways as they can.*

RIGHT: *One eco-friendly way to tour Cape Breton is the Celtic Shores Coastal Trail, which starts at the Causeway and finishes on the beach.*

The barrier bar across the pond has been enhanced and reinforced to serve as a breakwater protecting the photogenic, small fishing harbour — a popular destination for tourists seeking fresh seafood.

Visitors to Inverness appreciate how easy it is to get from the large parking lot directly onto the beach and then to remain under the watchful eyes of the lifeguards or stroll for great distances in soft, warm sand.

This barren exposure receives the full force of the les suête *winds, named by the resident Acadians for the southwest direction they come from.*

The ninety-two-kilometre Celtic Shores Coastal Trail brings avid cyclists from the Canso Causeway all the way to Inverness as part of the Trans Canada Trail. The beach is a perfect terminus for the nature trail, offering golden vistas where there was once just a grim horizon of pitheads and coal dust.

Shafts from the town's old mine run under the beach, following dark seams of coal well out under the gulf. They are flooded now. The mine is long closed, and the grimy dust that once blew around with the sand is gone. The only minerals left to gather are the specs of feldspar and quartzite washed down from Beaton's Mountain. They glisten brightly in the late afternoon sun.

Besides being warm and picturesque, the whole length of Inverness beach is littered with tumbled fragments of purple and maroon sandstone, mingled with bleached shells and multicoloured sea glass. It creates a huge, real-life Zen garden, like the ones found in trendy craft stores. Tiny sand sculptures, like parapets, attest to the constant sea breeze that impacts the scene.

It is a very exposed beach. Infamous *les suête* gales blow in from the southwest and have forced other sand-covered beaches along this coast to gradually retreat landward or disappear entirely. Inverness

The sun rises on a golfer's dream: a classic links fairway next to a rustic coastal panorama.

is less affected, partly because a thick layer of sea ice protects it from winter gales, but also because it shares its vulnerable piece of ocean frontage with a remarkable new development. While aggressive storms drive surging waves that could completely overwash its dune, this beach is protected from blowouts because it's backstopped by a golf course, Cabot Links.

Separated from rolling fairways, tees and sand traps by only a boardwalk along the hummocky dune ridge, the divide between the beach and the course actually marries the two landforms surprisingly well. Marram grass on the dune ridge stands face-to-face with the creeping red fescue of the large, rolling greens. They are unlikely to ever mingle, however, as each is perfectly suited to its side of the line.

Some say the three particular holes along the dune are the best run of fairways of the modern era, offering golfers intimate contact with the seaside, the mark of a true links course. Beachgoers near these holes share the pleasure of watching some of the world's best golfers, as well as celebrity players, the rich and the famous who fly in just to challenge this renowned layout.

INVERNESS

The smooth sandstone sits like paperweights, creating a landscape of mini-sandcastles.

Where there were once mine tailings, rusting machinery and rough roads, the brownfield site between the town and the beach has become one of the most environmentally friendly golf courses anywhere in the world. Thoughtful design has incorporated grasses that are suited to clay soils and the salty habitat. The cooler temperatures and lack of humidity reduce the need for pest management, and with very little irrigation required, the beach is spared a runoff of polluting chemicals.

One unexpected benefit witnessed on the beach since the golf course has been open is an increase in wildlife. Discouraged for many years by the noxious mix of coal dust and industrial debris, eagles, fox and deer have returned to the beach area.

Developers who want to sell expensive new homes build most of today's golf courses as incentives for people to buy into their subdivisions. The owners of Cabot Links took one look at the beauty of the beach in Inverness, however, and were convinced that it would sell their golf destination. They were right.

The partnership of the two attractions serves this former coal-mining centre well. The revitalized town is expecting steady growth of tourism, the likes of which the old miners could not have imagined a generation ago.

ASPY BAY

Cape Breton • 170 km north of Sydney

THE GALES OF THE NORTH ATLANTIC have been blowing into Aspy Bay for thousands of years. Arriving with the winds have come Norse adventurers, Basque whalers and the Italian master mariner Giovanni Caboto — or John Cabot as the English king called him when he granted him a commission to find a sea route to Asia. Cabot picked this sandy shoreline as his first landfall, on June 24, 1497. To another wandering sailor, it brought back visions of his home near Pic d'Aspe in the rugged Pyrenean mountains, and so it got its name. More recently, Scottish, English and French settlers have come here on purpose to make this bay their home.

Whatever the old country memories they leave behind, newcomers are welcomed to this bay by a string of four pearly beaches, eight kilometres long, which shelter a small harbour and fertile marshlands.

Seen from the deck of an approaching vessel, Aspy Bay resembles a huge bite taken out of this northern coast of Cape Breton. Over fifteen kilometres wide, headland-to-headland, it is a gap in the highlands that actually started three hundred and fifty million years ago as an ancient fault line. Geological records indicate that this is where two plates of the earth's crust were wrenched apart.

Halibut Head, Money Point and the other headlands of Aspy Bay are very resistant volcanic rock, so most of the sand found here has eroded from softer rock outcrops in the highlands. It arrives as gravel washed down through Big Intervale, the fault line

With the warm sand and a deep runnel to cross below, beachgoers leave their footwear high and dry.

that slid open and repositioned Cape Breton's two highest mountains by an additional fifteen metres. Coarse but colourful sand grains are deposited in fan-shaped fields at the mouth of the rivers. The dunes are almost always receiving a new supply of this sand, blown away then returned by the winds or carried in by surface currents that swirl around Aspy Bay, sometimes overwashing the dunes.

Called a bay mouth barrier, it is typical of sand accumulations all along North America's east coast, separating harbours and lagoons from the open sea.

Rivers discharge into these embayments and create estuaries full of brackish water, mud and silt.

For centuries, shrubs and marram grasses have covered the glittering bars of sand. Constant breezes challenge the vegetation, however, keeping it cropped low and forcing it to send taproots deep into the ground as anchors.

With the bay almost closed off, smaller streams like Wilkie Brook have to continuously carve new channels, runnels, between the dunes and through breaches in the barrier. To create the necessary

opening, the water just picks up the multicoloured grains of sediment and redeposits them a little farther down the bar.

The wave-cut plateau formed at the mouth of Wilkie Brook rests upon the gravel outwash from several North Mountain streams. There, a secluded jewel of a picnic park is the venue for summer family outings, plus the annual celebration of Cabot's landing. By taking the stairs leading to the beach, early morning walkers can easily spot the tracks of deer or other nocturnal scavengers, such as coyotes and raccoons, competing with gulls and eagles for carrion stranded on the shore.

Regularly studied by scientists and reported on by

ASPY BAY

ABOVE: *Rarely seen on the beach in daylight, local residents, like the raccoon pictured here, roam the dunes after everyone else has gone home.*

LEFT: *Often left off the traveller's itinerary, this eight-kilometre stretch of soft, copper-coloured sand is the longest beach in the province.*

earnest environmentalists, the beaches of Aspy Bay have received considerable attention as a model of the natural beach process, truly wild places where waves and currents move sand along the shore, between the dunes and out to the offshore bars. Their studies, however, have been measuring only the recent movements of Aspy Bay beaches. As a testament to nature's design, the beaches of Aspy Bay are the very same meeting place of land and sea they have always been. Academics have looked at the sculpting and scouring of these beaches in human terms. The real story of this place, though, is measured in the many thousands of years they have survived. The beaches of Aspy Bay are so old, they are timeless.

This classic barrier beach system is the fluid frontier between land and sea. The true way to take the measure of this place is not with a compass or a metre stick, but to walk the shoreline, barefoot. Let the cool waves wash the coarse sand between your toes.

It was the first place to be named by visitors from away. It is the best place for all beach lovers to visit.

25 NORTH BAY

Cape Breton • 125 km north of Sydney

THE LONG STRIP OF SOFT, DARK SAND in North Bay faces directly out to the open sea. Local beachgoers and visitors from the nearby picnic park know that it is most often protected from the stiff ocean breezes. North Atlantic swells never make it this far. What saves this beach and its cover of sand from being washed away, like so many others in Nova Scotia, is a pair of rocky headlands that reach far out to sea, plus several offshore shoals that sap the energy from any incoming tidal surge.

Like the cobble bar at Ingonish Beach on the south side of nearby Keltic Lodge, the beach at North Bay is unique. Technically described as a bayhead beach, it is a much longer strip of sand than Ingonish and a favoured stopover point during a Cape Breton vacation.

Clyburn Brook has long scoured the glacial till from the roughest parts of the highlands, delivering it to the shore on the north side of Middle Head. The broad meadow between the beach and the mouth of the brook is home to large patch of native thistle bushes, which is the likely origin of the local name for the beach, Thorne Beach. Deposits of so many different rock types from the Cape Breton Highlands constantly renew the supply of

RIGHT: *The shallow waters of North Bay prevent huge waves from reaching the shore and offer safe haven for the fishing fleets. Farther out, the shoals stretch for seventeen kilometres, supporting a thriving population of snow crab.*

BEACHES OF NOVA SCOTIA

ABOVE: *River valleys, once carved by glaciers, are washed clean of their sediments as Clyburn Brook tumbles through the roughest parts of the highlands. At its mouth, a meadow and the sandy shoreline provide an inviting contrast to the brook's rocky headwaters in the rugged upland gorges.*

RIGHT: *Several varieties of sea rocket flourish along the Atlantic's exposed shores. Early pioneers may have used the succulent leaves and stems in salads.*

OPPOSITE PAGE: *Unless the sea is riled up by extremely strong onshore winds, the waves gently lap the shore by the time they reach the beach.*

multicoloured minerals on the beach. Feldspar, quartz and limestone crystals, chipped from rocks after years of tumbling downstream, mix with flotsam and driftwood in a natural seaside tableau.

It is the enduring struggle between the forces of the sea and the stubbornness of the land that plays out so dramatically at the shoreline. All along this coast is the evidence that nothing lasts forever.

Winter will cloak the beach with a blanket of ice, shielding it from the worst of the weather. When the ice begins to break up in the spring and the occasional storm surge returns to worry this beach, the leading edge of the forest cover always loses some ground, literally. Over time, the natural process will have the effect of moving the beach steadily landward.

With the gradual erosion of the shoreline, trees

NORTH BAY

like the white spruce and bayberry will be felled. Coastal bird habitats and woodland cover for small animals will also be reduced. Whether a textbook example of the perils of rising sea level or just the cyclical struggle between the forces in the natural world, low dunes and shore ridges have been eroding here for millennia. As latecomers to this land, local residents recognize these transformations as part of larger, long-term cycles and have begun to adapt their own land use practices to account for nature's unstoppable drive.

INGONISH

Cape Breton • 115 km north of Sydney

THE INGONISH AREA IS KNOWN FOR several scenic attractions and for both summer and winter recreation. It is also one of the few tourism sites with two distinctly different beaches. Middle Head neatly separates the two parts of Ingonish Bay like the index finger of a king, and hosts Keltic Lodge as a splendid crown. Lying in repose below its cliffs is the barrier bar, Ingonish Beach.

The Cabot Trail is widely renowned as a destination, and coach tours often include a stop at this stony dyke across South Bay where visitors are fascinated by the high, wide stretch of cobblestones. This classic barrier beach has long separated the breezy Atlantic Ocean from the warm waters of Freshwater Lake.

The stones on the bar are tossed constantly by the sea, rounding their edges and giving them a polished evenness, not often found in nature. Parks Canada offers a number of services and facilities here, like boardwalks over the loosely packed stones for easier access to the supervised section of the beach. Beachgoers love the sandy beach, but except for the hearty few, most appreciate the more swimmable waters in the lake behind.

There is also a short but steep climb to a lookout with a panoramic view of the bay at the north end of the beach. As a word of caution, however, poison ivy near the lookout requires special care.

The fast-flowing waters of Ingonish River, fed by excessive snowfalls in the highlands, tumble the ancient granite rocks, and even some long-dead relics

INGONISH

Though pretty from a distance, the well-rounded stones are unstable, making them very hard to walk over.

BEACHES OF NOVA SCOTIA

As counterproductive as it may sound, some fish must come right up on shore to breed. Capelin swarm the coarse sand beaches of eastern Cape Breton in mid-summer to briefly spawn and hopefully catch a returning wave back to the sea.

ABOVE: *Adding to the pleasant scene, the Atlantic Ocean water in late summer is almost warm.*

LEFT: *Released from their mountain home, the feldspar, quartz and limestone crystals make a multicoloured beach sand.*

of forests forgotten, down to the shore. Rounded and reduced in size by the constant knocking about, both eventually fetch up on the exposed sand of South Bay and are stacked neatly into place by the ocean waves. The stones roll back and forth, while the passing currents carry the sandy fragments of quartz, feldspar and limestone away to offshore sandbanks.

In the summer, the beach does receive a broad line of soft, pink sand thanks to the currents that carry the colourful crystals back in from the offshore banks. Adding to the tourism values, for a short few weeks, the temperature of the seawater approaches pleasant — *almost*. Like the warmer water, however, the sand lasts only until the fall storms whisk it and the colourful autumn leaves away.

In winter, the snowy white beach centres the view the skiers get from the chairlift atop Cape Smokey.

27 PONDVILLE

Cape Breton • 55 km southeast of Canso Causeway

MANY OF THE BEACHES ON THE Atlantic Coast are a glittering gold colour. Rays of sunshine bounce off the angular grains of quartz and mica that make up the bulk of their sand. The wide, flat beach at Pondville, on Cape Breton's Isle Madame, is a steel grey, like the glacial till left there by the receding of the last ice cap.

The shoreline boasts two durable outcrops of ancient granite and quartzite, which were not ground down by the great masses of ice. Petit Nez to the north and Gros Nez to the south reach six kilometres out into the open ocean. At first glance, the formidable old cliffs do look like stony noses.

Big swells, however, rarely reach past them this far into the bay. The waters are shallow between the two headlands, marked by numerous sandbars and covered by thick beds of rockweed. Waves that do overcome the natural reefs wash into the beach carrying only the finer grains of silt, mixed with fragments of seaweeds and thousands of tiny sea creatures. In this generally low-energy environment, the beach remains fairly flat and the sand packs solidly between the crevasses of durable granite, not quite covering the topmost ridges.

The people who found shelter here in the bay, between the rocky crags, were as resilient as that granite.

Over two hundred and fifty years ago, Isle Madame became a new frontier for displaced families of several cultures. Indeed, all of Cape Breton

The marsh held back by this barrier beach marks the transition between the land-based agriculture, thought by the early settlers to be their main ambition, and the productive seashore where an abundant fishery has grown to support many new generations.

offered a well-earned peace for demobilized British soldiers and displaced Acadians, for uprooted Irish serfs and New Englanders who preferred self-exile rather than new rulers. These were the pioneers who joined the natives on this exposed island.

Their chosen frontier was this very beachfront, in the sheltered cove at the end of the Bay of Rocks. Most of the Irish peasants who gave it its name were first interested in the farming potential of their newly granted pasture land. Although many of the shades of green here are similar to the farmlands they left behind, even they had to recognize the greater abundance the sea and shore provided. Pondville, then and now, borders a bountiful fishery in the bay. One unheralded pleasure has been the discovery that the sand is a perfect consistency for building sandcastles, leaving creative impressions or just upside-down footprints.

Scattered along the sandy beach is the typical debris of rich inshore fishing grounds. With every new tide, remnants of discarded rock crab shells, slipper limpets picked clean by hungry gulls and a variety of seaweeds are all scattered about. Mussels, clams, periwinkles and the rockweed itself provided a year-round harvest along these shores, which are rarely covered by winter ice. The flotsam is also a good indication that a productive lobster ground lies under the nearshore waters.

BEACHES OF NOVA SCOTIA

ABOVE: *The sand and silt mixture here is a unique consistency, resisting the sea breezes and showing off its upside-down footprints.*

RIGHT: *Scattered along the sand is the typical debris of seacoast villages. Remnants of rock crab along with slipper limpets, picked clean, have been scattered by hungry gulls.*

The pond at Pondville, held back by the sandy beach, has grown thick with saltwater grasses. It used to support hay fields, shellfish beds and colonies of shorebirds — all of which, in turn, fed the pioneer settlers. They even valued the salt hay as insulation around the footings of their classic design Acadian cottages. Nearby they found basins of thick, black peat to carry back to their winter hearth.

An outcrop of very old and hard rocks lies along the northern part of this cove. Most of the area has a silt and shale foundation, though, giving the beach a soft grey coating of sand.

A steady flow of fresh water drained through the fertile Barachois Pond and nourished abundant quantities of small plants and animals. The tide flowed in and out of the narrow inlet, carrying all that naturally organic nutrition out onto the nearby shellfish spawning beds. It is no wonder local fishermen established a wharf at the mouth of the creek, securing their small boats as near as possible to the bountiful inshore fishing grounds.

As peaceful as it seems, the beach is always in transition. At times of severe weather, surf breaks on this beach. Winter winds attack the dune, and the sands are relocated more often than even the frequent visitors notice. Actually, as Pondville absorbs these natural forces, it is performing the acts of a healthy beach. It remains the frontier between the land and the sea: ever changing, but never seeming to.

One of those changes happened too long ago for most people to remember. These shores were once a haven for great herds of walrus. Their niche is now occupied by a variety of seals and smaller marine mammals. From the beach, several large bobbing heads can often be seen scrounging for abundant seafood, looking for a mate or just being curious about who is watching them.

THE BEACHES ACT

THE GOVERNMENT IN NOVA SCOTIA HAS officially enshrined the beaches of Nova Scotia and dedicated them in perpetuity for the "benefit, education and enjoyment of present and future generations of Nova Scotians."

In 1989, an act of the legislature provided for "the protection of beaches and associated dune systems as significant and sensitive environmental and recreational resources." The act also placed most beach activities under regulation, so that driving on a beach would not be permitted, as well as unregulated mining of the sand and gravel, among others.

Just what is a beach and how much land is to be considered a part of it under the legislation can be determined as the government sees fit. This has positively influenced land-use decisions in hundreds of locations and has been responsible for bringing ninety-two specific sites under designation.

ACKNOWLEDGEMENTS

THE AUTHORS ARE GRATEFUL TO THE MANY people who shared their knowledge and expertise about beaches. Rick Welsford, Joliene Stockley and Piet Mars were very willing to share their appreciation for the hidden values which all beaches add to our lives. Harold Carroll, Art Lynds and Jonah Bernstein added their insight and expertise as well. They are also grateful to Helen and Ley Chapman and Allen Chapman for their unwavering support throughout this project.

INDEX
PHOTOGRAPHIC REFERENCES APPEAR IN BOLD.

A
Acadia, 93, 112, **133**, 151, 152
agate, 83, 93
amethyst, 83, 93
anemones, 91
Atlantic brant, 70
Atlantic Flyway, 70, 98
Avon River, 84,

B
back beach, 42, 46, **54**, 63, 129
balsam fir, 12
Barachois Pond, 108, 153
barnacles, 80, 91
barrier bar (beach), 30–1, 52, 74, 110, 124, **126**, 132, 137, 139, 146, **151**
Barrington Bay, 62, 66
basalt, 78, 80
Basque, 136
Bay of Fundy, 5, **73**, 74, 78, 80, 82, 83, 84, 90, 92
bayberry, 145
bayhead beach, 140
beach cusps, **38**, 41, 45
beach pea, 30
beach rose, 123
Beaches Act, 7, 154
Beaton's Mountain, 133
Beatty Marsh, 94, 96, 97
Big Intervale, 136
Boardwalk Beach, 120–1
    See Also Shipping Point
boardwalks, elevated, 65, **74**, **77**, 108
bog cranberry, 122–3
Bridgewater, 48

Brule Shore, 98
bulrushes, 98

C
Cabbot, John, *see* Giovanni Caboto
Cabot Links golf course, 134–5
Cabot Trail, 146
Canso Causeway, 133, 150
Cape John, 99
Cape Sable Island, 62, **65**, 66
Cape Smokey, 149
Cape Split, 78
capelin, 14, **16**, 148
Carboniferous Period, **87**, **88**
cattails, 98
Celtic Shores Coastal Trail, 132, 133
clams, 33, 35, **74**, 95, 96, 114, 121, 151
clam digging, 57, 74
clothing-free option, 41
clubmoss, 17, 36
Clyburn Brook, 140, **142**
coal mines, 125, 135
coastal barren, 12, 36
Cole Harbour, 30
Cole Harbour Marsh, 31
common tern, 14, **16**
copepod, 80
Cow Bay, 5, 7
crowberry, 17, 36

D
deer, **126**, 135, 138
dogwood, 17, 36
dolphins, 121

dunes, 18, 21, 28, 31, **38**, 69, 70, 75, 95, 111, 122–3, 124, 126, **139**, 153
    formation, 19, 22, 23, 40, 46, 49–51, 55, 62, 66, 72, **73**, 74, 94, 98, 102, 104–5, 108, 110, 120, 121, 137
    succession, 109
Dunn's Beach, 108

E
eagles, **82**, 112, 135, 138
Eastern Passage, 5
eelgrass, 41, 70
Evangeline Beach, 80–1, 82

F
feldspar, 40, 93, 125, 140, 149
Flying Point, 22
Fortress Louisbourg, 120
fossils, **87**, **88**, 93
Freshwater Lake, 146

G
gaspereau, 32
Giovanni Caboto (John Cabot), 136, 138
Glooscap, 78, 83
granite, 36, 40, **58**, 146, 150
great black-back gull, **129**
Green Bay, 52, **53**, 55, 57
greywacke, 66
Gros Nez, 150
Gypsum, 83, 84

H
Half Island Point, **28**

INDEX

Halibut Head, 136
Hantsport, 84
herons, 32, **51**, 77, 112
herring gulls, 91, **129**
highlands
    Cape Breton Highlands, 130, 136, 140, **142**, 146
    West Mabou Highlands, 124
Hirtles Pond, 49
Hubbards, 42
Hummocks, 66, 90

I

ice age, 58, 70, 80, 86, 102
Important Bird Areas, 70
Ingonish River, 146
Ipswich sparrow, 33, 34
Isle Madame, 150

J

Jackies Island, 60
Japanese rose, 123
jasper, 83
jellyfish, 14–7, 58

K

Keltic Lodge, 140, 146
Kings Head, 102
Kingsburg Peninsula, 48, 49

L

LaHave, 48, 55
lamb's quarter, **118**, 119
leatherback turtle, 14, 16, **17**
lepidodendron, **87**
les suête winds, **130**, 133
limestone coral, 84, 144, 149
Little Harbour, 102
longshore drift, 45, 47, 62, 104
Lunenburg, 48

M

Macdonald Bridge, 5
MacInnis Pond, 95
MacIssac's Pond, 130
mackerel, 29
Malagash Point, 94
marshes, **33**, **51**, **54**, 66, 108
Massacre Island, 60
Melmerby (vessel), 107
Mi'kmaq, 78, 83, 111, 112
mica, 40, 150
Middle Head, 140, 146
Minas Basin, 78, 92
mollusk, 116
Money Point, 136
Monks Head, 108
morning glory, 30, **34**
mud shrimp, 80, 81, **83**, 91
mummichogs, 95, 32
muskrats, 98

N

Norse adventurers, 136
North Mountain, 78, 138
Northumberland Strait, 94, 95, 102, 107, **128**

O

offshore bars, **41**, 72, 124, 139, 140
osprey, 29, **82**, 112

P

Parrsboro, 78
Pennant Point, 36, 38
Pensey Head, 30
peregrine falcon, 82
periwinkle, 33, 116, 151
Petit Nez, 150
Petit Riviere, 52
Pic d'Aspe, 136
pilot whales, 121
piping plovers, 65, 70, 111–2, 113
pitcher plant, 17
Point Enrage, 48
porpoise, 121
Psyche Cove, 12

Q

quahogs, 96
quartzite, 135, 150

R

raccoons, 91, 118, 138
red maple, 77
red-winged blackbird, 98, **99**
Riverport, 48
rock crabs, 151, **152**
rockweed, 80, 150, 151
rogue waves, 29
Romkey Pond, 48
Roy Island, 102

S

sea levels, 46, 47, 57, **70**, 71, **78**, 80, 86, 109, 145
salt flat, 70
salt marsh, 31, **33**, **34**, **51**, 54, 72, 74, 76, **86**, 87, 94, 97, 108, 124, 129
samphire greens, 16, 30
Samuel de Champlain, 52
sand dollar, 18, 30, 58
sandpiper, 18, **19**, 32, 65, 80–1, 91, **97**, 99, **100**
sandstone, 52, 66, 78, 90, 92–3, 97, 102, 130, 133, **135**
sandwort, 62
Savannah sparrow, **33**, 34
Scotia Square, 7
seashore buttercup, 30, **33**
seaweed, 48, 53, 55, 57, 80, 94, 101, **114**, 118, 150, 151
sandpiper, 18, **19**, 65, 80–1, 91, **97**, 99, **100**

September Storm Surf Classic, 26
shale, 66, 78, 84–8, 153
sheep laurel, 122
shingle bar, 84
Shipping Point, 120–1
        See Also Boardwalk Beach
siltstone, 84, 86
Silver Sands Beach, 5, 7
slipper limpets, 151, **152**
smelt, 32
snipe, **76**, 77
Spectacle Island, 60
sphagnum moss, 122
spotted sandpiper, 99, **100**
spruce, 12, 30
    black spruce, 58, 76
    scrub spruce, **139**
    white spruce, 145
spume, 125, 129
St. Margarets Bay, 42, 44, 47
St. Mary's Bay, 72
starfish, 91
storm surges, 23, 28, 46, 49, 54, 70, 100 , 102, 114, , 120, 144
strand line, 118, 119
surfing, 18, 24, 26, 76, 130
swash zone, 100, 105–6

T

tamarack, 76
Thorne Beach, 140
tidal pools, **51**, 80, 90–1, 94, 95, **96**, 97
tombolo, 55, 102
Trans Canada Trail, 133
Treen Bluff, 94
Triassic Period, 93
trilobite, 101
Turkey Dip, 26

U

United Empire Loyalist, **13**
urban renewal, 5, 7

W

walrus, 153
water temperature, 24, 41, 52, 63, 74, 96, 113, 121, 130, 149
wetland, 77, 94–7, 98
wild rose bush, 30, 74, 109, 123
Wilkie Brook, 137, 138
willet, 77
Wobamkek Beach, 60

Y

yarrow, 74
yellowlegs, 32, **54**

Z

zostera, 90